The Journals of
Walter White

Assistant Secretary of the Royal Society

WALTER WHITE

CAMBRIDGE
UNIVERSITY PRESS

CAMBRIDGE UNIVERSITY PRESS

Cambridge, New York, Melbourne, Madrid, Cape Town,
Singapore, São Paolo, Delhi, Mexico City

Published in the United States of America by Cambridge University Press, New York

www.cambridge.org
Information on this title: www.cambridge.org/9781108045131

© in this compilation Cambridge University Press 2012

This edition first published 1898
This digitally printed version 2012

ISBN 978-1-108-04513-1 Paperback

THE JOURNALS

OF

WALTER WHITE

Walter White.

THE JOURNALS

OF

WALTER WHITE

ASSISTANT SECRETARY OF THE ROYAL SOCIETY

WITH A PREFACE BY HIS BROTHER

WILLIAM WHITE

AND

A PORTRAIT IN PHOTOGRAVURE

LONDON

CHAPMAN AND HALL, Ld.

1898

RICHARD CLAY & SONS, LIMITED,
LONDON & BUNGAY.

PREFACE

WALTER WHITE'S Diary owes its interest to various sources. His long employment at the Royal Society brought him in contact with many eminent persons during a long stretch of years, from 1844 onwards; and the numerous topics, literary as well as scientific, suggested by these meetings, are amply dealt with in the journal from which this volume is compiled. Tennyson and Carlyle, Faraday and Tyndall, Grove and Huxley, Wheatstone and Airy, are names which will at once occur to the memory of the reader of this volume. Their number might be added to from all quarters; but it is necessary here only to point out the salient features of the Diary, and not to summarize its contents.

Another element of interest is associated with the journal from the fact that it records the life—almost from day to day—of a man who strove unsuccessfully to earn a satisfactory livelihood as a cabinet-maker, who educated himself far above the level of most gentlemen of his day, and who eventually obtained a position which was not only congenial to his tastes, but useful to those with whom he was associated. The contrast shown in these pages between the workman of early days, and the Assistant Secretary of the Royal Society of later years, is one of no small interest and attraction. In 1844 Walter White became Sub-Librarian to the Royal Society, then in Somerset House. Within ten years he was promoted to the post of Assistant Secretary. In both capacities he undoubtedly acquired the faculty of making himself extremely useful to the Fellows of the Royal Society. He had long been able to read, write, and speak French and German, and had some knowledge of other continental languages. His general education, mainly self-acquired, was thorough and practical. He was accurate, painstaking, and receptive. He saw when a proper chance in life was offered him, how it could best be used to the advantage of himself and of others. And every characteristic here indicated is traceable in his Diary. The original handwriting of the diarist has been carefully copied and followed. The journal itself is contained in four octavo manuscript volumes, closely written in a small, and not always clear hand. It has been thought better to give mistakes where they occur, in the form in which the diarist has recorded his thoughts.

Passages relating to the late Lord Tennyson have been selected and arranged in chronological order as a separate chapter. It is evident that the two men had something in common, at all events for a time. No effort is made to add to the interest which the Diary possesses, from extraneous sources, except perhaps in printing the letters to which Walter White owed his introduction to his predecessor in the office of Assistant Secretary of the Royal Society, and it has rarely been found necessary to burden the pages of the volume with matter that is not contained in the Diary itself.

The foregoing remarks fully express the reasons which have induced me to consent to the publication of selections from my brother's Diaries. As the eldest son of a large family, he was looked up to by his juniors with some deference and admiration. Always with a strong sense of duty, he followed his daily calling with conscientious diligence while he engaged with much zest in the recreations and amusements common to youth and early manhood, in all he did manifesting a lively humour.

My brother was a diligent reader and student from his school-boy days, and early committed his thoughts to writing, both in prose and verse; in the latter often developing the humorous in relation to the incidents of family life, but often also to express his filial regard for parents and home. Some of these rhymes, on varied subjects, appeared in several periodical publications, and were afterwards published in a separate volume.

As will be seen in the earlier pages of the Diary, my brother's aspirations were never fully satisfied until he found congenial employment in the service of the Royal Society, in which he was engaged for upwards of forty years, with so much satisfaction to himself and his principals.

WILLIAM WHITE.

Edgbaston, Birmingham,
October 1897.

CONTENTS

THE JOURNALS

OF

WALTER WHITE

CHAPTER I

EARLY YEARS, 1811—1844

THE writer of these diaries was born at Reading on April 23, 1811, of parents whose intellects and education were far superior to those of most tradespeople early in this century. In religion a Methodist and by trade a cabinet-maker, the father brought up his son to follow in his footsteps. Quoting Walter White's recollections of his early childhood, written in 1843, we find he says—

My earliest recollections go back to the time when I was about three or four years old, and the only circumstances I remember of this time is repeating a hymn sitting on my father's knee

—this was a favourite fireside occupation on the Sabbath afternoon. I have some dim remembrance of being praised for my repetition. The hymn was, I believe, one of Miss Taylor's on Creation; very simple, and well adapted to the mind of a child; one line of it I have never forgotten from that time to the present, which arises, I suppose, from its flowing musically and pleasantly off the tongue—" The rivers and the rills." . . . My school reminiscences of this period are but few and uninteresting. I went with my brothers and sisters to a school for little boys and girls, kept by a maiden aunt of ours, who is yet alive and to whom we are indebted for initiation into the mysteries of the alphabet, and *ba, be, bi, bo, bu.* My memory retains no impression of the manner in which the victory over these battalions of difficulty was achieved. But after the completion of the initiatory process, the safe passage of the *pons asinorum,* I remember that I could spell better than any other of my school-fellows and was always ready with my answers to the Catechism, in which we were exercised regularly once a week. I was fond of reading and anxious for instruction, but being of a fearful and timorous disposition, and withal not a little dull and

stupid, I was universally considered a dunce.
My ignorance of everything except the mere
words of my book was complete, still I was
always at the head of the Testament or reading
class, till one fatal day, seduced by a bribe from
a boy who was my senior by two or three years,
and whose knowledge of general matters was as
great as mine was small, seduced I say by this
boy into a betrayal of my position, I suffered my
eye to wander from my book, became confused
when my turn came. My betrayer took advantage of this, read my verse, and walked *up* into
my place while I *descended* into his. I shall
never forget the agony of mind this caused me,
how bitterly I repented of my folly, and determined to hold myself constantly on the watch
to recover the position I had so foolishly lost.
Young as I was I had the sense to understand
the triumph of my opponent was not a fair one,
one for which I was indebted to my weakness
rather than to his own talent. I do not remember that I ever again occupied my former
position at the head of the class.

Sewing was a pursuit I sometimes followed
while at this school, my performances in this
feminine art were very well and neatly executed.
I had a natural love of neatness and order and

was not a little pleased when I carried home a bag as the fruits of my industry. This love of order has remained with me, an untidy room was and is my abhorrence. When a boy I have often seized an opportunity when the rest of the family were absent to put the room where they had been sitting in complete order, and then disappear that their surprise might be the more complete on their return. I have often thought it strange that with all this love of order, I have never been able to arrange my thoughts in a clear and definite manner. In an argument as in reading I am very apt to lose sight of what has preceded and thus become confused and barren. I was at this time a most egregious coward, for which base quality I think I am partially if not wholly indebted to the advice of my mother. Her advice was always to run away from a dog, and if a boy met and assaulted me in the street always to take to my heels, never to think of defending my own rights but to submit to anything rather than to fight. "It is better," she would say, "to run a mile than to fight a minute." The consequence of this was that I became mischievous; it was not in the nature of boyhood to bear everything without attempting in some sort at retaliation. I therefore sought all opportunities of tormenting persecutors where I could do so without injury to myself, but in this I lost my self-respect. I despised myself for not being bold enough for a fair stand up fight. Could I have done this as I often most earnestly longed to do, I should have acquired greater firmness and independence of character and have entertained juster views of myself and those around me. I have often had to deplore this timidity, for when I have been on country walks and the footpath has led through a farm yard where there was a dog I have either made a great detour or stopped short in my walk, though a beautiful view perhaps lay beyond and which was the object of my stroll. A shower of rain would at the time of which I now write have put me into an agony of fear, from some unaccountable cause the sight of rain excited in me the deepest turns, and if I were indoors no inducement could tempt me out. After I had got over this fear it clung to me in another form which was the dread of bathing.

He resided at Reading with his parents, doing upholsterers' and cabinet-makers' work until October 1830, when he seems to have undertaken one of the tours he so much loved to different parts of England at first, and later on extending from the United Kingdom to different parts of the Continent. In 1830 he went to Derbyshire and Staffordshire, where he married at the age of nineteen, the only entry on the subject being in French, "J'allais à M . . . et je fis une heureuse alliance." He writes—

"Jamais je n'ai tant pensé, tant existé, tant vecu, tant été moi, si j'ose ainsi dire, que dans les voyages que j'ai fait seul et au pied. La marche a quelque chose qui avive et anime mes idées. Je ne puis presque penser quand je reste en place, il faut que mon corps soit en branlé pour y mettre mon esprit. La vue de la compagne, la succession des aspects agréable, le grand air, le grand appetit, la bonne santé que je gagne en marchant, l'eloignement de tout ce qui me fait sentir ma dependence, le tout qui me rappelle à ma situation, tout cela degages mon âme, me donne une plus grand audace de penser."

As early as in the beginning of 1833 we find him writing (with some anticipation of the last three years of his life)—

I find that much valuable time has been lost and misspent, many fond anticipations have been destroyed, and many schemes have been frustrated, many acquirements that I might have gained I have not gained. Let this not be the case at the end of my next book. May I become more active and less selfish, may I acquire habits of study and firm application. Here I am, in the possession of a moderate income, striving to confine my expenses within it, and to enjoy as many as possible of the comforts of life. I believe that I am happy and contented . . . and I hope to train up my family in a manner at once parental and creditable. What my income or circumstances may be at the ending of the next book the Searcher of hearts alone knows. May I be able to review the time elapsed with more real satisfaction than at present.

In February 1833, he and his family had settled down near his parents in Reading, where he was busily occupied in making furniture during the day and improving his mind at night. On March 1, 1833, he writes—

Read Latin; the more I study this language the more I like it, and the more my desire to know it increases.

1833.

March 2. Working at wardrobe, bought watch. In the evening got my usual monthly publications, 'Chambers' Journal' and 'The Penny Magazine,' peculiarly interesting. Commenced tabling ' Companion to the newspaper.'

March 5. Finished wardrobe and getting pianoforte in at window in Castle Street, a very troublesome, fatiguing job. Not able to study so read ' Rivals' in the evening, an interesting work.

March 6. Made deal picture frames, finished loo table, and began two basin stands. Working late in the evening, began chest of drawers. Read Latin.

March 13. Began wings to wardrobe and two Loo tables. Paid part of bill to Mr. Jacobs and Page, nothing like being out of debt.

March 15. Working part of day at Loo tables, cogitating during the day on my tale of the Puritans, hope it will succeed, courage ! working late in the evenings.

March 17. Mr. Rowe preached from this text,

" Let the words of my mouth." Rained fast on coming out of chapel. Mem. : very vulgar to run home.

March 18. Received parcel containing Latin books from Underwood with a very witty letter.

March 20. Feel much inclined to continue writing and to finish my tale which I hope may prove successful.

March 21. Worked hard all day. It will not do to indulge idle habits. Read 'New Monthly' after tea, some of the articles good, wish I could write such.

March 22. At 12 o'clock assisted in removing a pianoforte from Mr. Lomers' to Caversham, succeeded to perfection, no damage. Very hungry and no food.

March 25. Marie wrote letter to Smith and Elder, hope to know the fate of manuscript before end of week.

March 26. Cogitating much on my tale during the day, must set myself assiduously to work to finish it.

April 1. Working at wardrobe, application in its strictest sense essentially necessary. Got magazines in the evening, read them.

April 2. Working at wardrobe, fine working weather. Too much inclined to be idle, of no

use to make resolutions without strictly adhering
to them.

April 10. Anticipating perhaps too much from
my tale but hope will linger.

And on the 13th, the entry "Manuscript
returned from London." But this does not seem
to have deterred him, as two days later he
mentions that he is again writing, and on the
23rd he says—

Reflecting on wealth and happiness. Remain
satisfied with my lot and envy not others. Strive
to retain this feeling.

May 27. Wonder if I shall become religious
or learned, surely not without trying; may
Providence assist me.

May 31. Musing on poetry and prose during
the day, felt many ambitious thoughts arise but
only moderate ones.

June 1. Felt in quite a poetical mood in the
afternoon, but had not time to indulge or to
make use of it.

June 15. I should like to be a preacher, that
is a real good one. Bought first number of
' History of America.'

Two days afterwards an accident befell him
and he cut the top off his finger, which prevented

his working, but did not interfere with his writing. On June 28 he writes—

Cannot help fancying my prospects rather gloomy, but must strive to be contented.

June 21. Walked with bills. . . Read a series of articles in 'The Tourist,' lent me by G. Rickman, relative to the effect produced on the minds of men by the study of the Classics, cannot agree with all he says, a knowledge of the ancients certainly does improve a man.

June 22. Walked out with bills . . . and in, bought number of 'History of America.' Have in one sense lost a whole week, though I hope I improve my mind. Knowledge I think is all I want.

June 25. Rose at usual time and much pleased to go to work again. Continued with hat-stand and chiffoniers. Dr. Clarke's opinion, "a late morning student is a lazy one, and will rarely make a true scholar, and he who sits up late at night not only burns his life's candle at both ends but puts a red-hot poker to the middle." I should like to act oppositely to the above, but fear my labours will prevent. Read 'Christian Advocate.'

July 1. Reflected much on the means of supporting my family, no very bright prospect.

July 3. Must work hard in spite of all, and do my best while relying on Providence for succour.

July 4. Determined should nothing better offer to go to America, for there only can I see reasonable hope of providing for my family. Spoke to mother on the subject, must work diligently, be economical and gain sufficient money.

July 5. Worked heartily all day under the idea of working out my better fortune. There must be no waverings now, it must be decided. Talked with father who gave me encouragement on the matter. Wrote all the evening, think of departing March or April 1834.

July 6. Worked heartily again, got the first sixpence to-day.

July 7. Heard Mr. Rowe from "All Scripture is given by inspiration of God." It has quite convinced me of the truth of the Bible which I was before inclined to doubt. His arguments were lucid and forcible.

July 8. Talking with father about the Puritans. Then Wesley, Cobbett, and Tom Scarlett and twenty other subjects. Worked willingly. . . . In evening commenced deposits in Saving Bank.

At this time White seemed trying every sort of employment, for his diaries for the rest of the year 1833 are but records of "drawing in the evening," writing, poetry, playing the violin, book-binding, reading, upholstery. Sermons heard at chapel every Sunday and thoughts of migrating to America as well as doubts as to whether he shall ever get there recur frequently. On January 22, 1834, he writes—

Feel anxious with regard to emigration, will it or will it not arrive? Dieu sait, however I shall try—lottery.

A few days later—

Infantine confidence a pleasing theme for a parent to dwell on. What a blessed thing is pure confidence where no vexatious doubts or fears intrude to mar happiness.

A month later are entries—

Must try for immediate emigration. Talked to father on emigration. Il faut essayer quelque chose que puis je faire pour aller en Amerique. Think of removing.

After trying fruitlessly for work near home and in London he returned to Reading to pack up and sail with his family for New York in April 1834. Here he obtained employment with Messrs. Copcutt and Son, the elder of whom had been in the employ of Walter White's father at Reading twenty years previously, and after emigrating had prospered and made a large fortune.

He found the alternations of successive heat and cold very trying at first, but struggled on at cabinet-making as well as his writing, as there are entries in his diary of "Went to Publishers several times." The next year, 1835, he seems to be more acclimatized as he remarks—

Do not feel the cold so much now as last time.

For a time "Work is brisk," and then come strikes in the trade.

1835.

March 31. Much discussion in shop, relative to the turn-out; refused to join, not in accordance with my principles.

April 1. Shop rather lonesome owing to absence of ouvriers, gives me time for reflection and meditation. May I learn to improve my time, and in becoming wise become better.

April 3. Mutual esteem and confidence between man and wife is surely one of the greatest of earthly blessings, and yet how little pains we take to secure them, when by striving to bear and forbear all might go well.

But very shortly he writes, February 3, 1836—
Talked of going to Buenos Ayres, do not like these severe winters.

And on February 13—
Went to enquire about Buenos Ayres.

Soon come letters—
From John wishing for our return (to England).

And on April 21—

Feel impatient to know whether we return to England or not. Perhaps it will be more to our advantage to remain where we are.

And the next day—

Have thoughts in case of staying here, of going into the country (Utica) to work, as it will doubtless be conducive to the health of the family.

1836.

May 20. Letter from father; wishes for our return, which we have planned for June 10, if all goes well.

June 1. Much surprised and pleased to see James just arrived from England.

But work was slack and Walter White had to look for it outside New York, first at Albany and then at Troy. On June 17 we find—

No luck in Troy, left at noon. Back to Albany, must take to the bench again.

June 22. Arrived in Poughkeepsie at 2 p.m., found work immediately. No lodgings, went to boarding-house. Maria sat in the dock for two hours, while I was searching, full of distress and trouble of mind.

August 14. Went to booksellers, great anticipations respecting riches. Prepared for upholstery work.

September 22. Rather unsettled in mind in consequence of thinking of change of occupation.

c

September 26. Quite ill au soir. Religion's powerful aid comes in to check the bold approach of sin.

September 27. Poorly all day.

> For I shall never be at rest
> Till by the daising turf I'm prest.

October 5. My thoughts continually directed towards finding some mode of occupation and bettering our condition. Think I should like teaching best if a fair start were made.

On October 13 he arrived in Schenectady, and—

Saw James. Dull prospects, must try hard, sink or swim. Wrote lecture.

October 15. Sixteen miles in fifty minutes, poor looking place. Made arrangement for delivering lecture [on learning French on the then popular Hamiltonian system] on Tuesday evening next. Permission to use Presbyterian Church.

October 18. Distributed circulars, and prepared for lecture, feel rather nervous on the subject. However succeeded better than I thought for at 7 p.m. When I first rose to speak felt like ten thousand muscles pulling my lips, did not know the sound of my own voice. Audience about fifty in number.

October 20. Began two [French] pupils au soir.

October 21. Some hopes excited from an interview with the students of the Academy.

October 26. Heard declamations of students of the Academy.

November 7. Not so many pupils as I expected, must not despair.

November 11. Dispirited with ill luck.

November 12. Shall I quit teaching and give up my ambitious hopes?

November 26. Home once more. Pleased to see progress of children; yet some happiness.

November 29. No work to be had, must try teaching again.

December 15. What is more unbearable than suspense? How ennuyant to be kept in ignorance. Sometimes think that I do not trust sufficiently to myself. Must improve the time when it does arrive.

December 16. The consciousness of debt incurred does not tend to cheer one. Must however try to keep my courage from sinking.

December 20. Thought of going to New York but cannot find the means. Hard case to want a dollar.

1837.

For the next half-year there are few records, excepting of books read and studies, and hopes and fears for the future. On August 30 he writes—

The few moments snatched from labour to devote to literature I esteem a treasure; they are like the green oasis in the desert to the weary traveller. When *will* thought become eloquent, and my brain like a gushing stream?

September 4. Oh, that my dull imagination would wake up and embody the dreamy fancies which float across my brain, that tipped with Promethean fire my pen might animate all it touched.

September 23. Feel my mind constantly full of poetry, but cannot convey in expressive language my dreamy fancies, perhaps after all I am no poet.

September 24. Do not believe that selfishness forms the basis of all human motives. Cannot believe in future nonentity.

October 1. Too much egoism in my nature, must sacrifice my own feelings for the welfare of my family.

November 30. Cannot divest myself of the idea that work is a surer means of advancement in life than teaching. How one clings to that which one has been accustomed to for years.

December 3. Read criticism on my poetry in 'The Mirror,' a spur to future exertion. Also an article on habit, setting forth that a man may by steady practice attain eminence in any profession. May I prove it by industry and success.

December 9. Feel depressed at my lack of skill in poetry. Oh that I could express all I feel.

December 10. How pleasant are the occupations of the mind. How delightful to peruse the writings of wiser heads than one's own.

December 11. Working late at book-case. Feel a pride in thus creating around us many sources of comfort and enjoyment. How much of the ills of life might be avoided if we would but try.

1838.

February 13. Talk and think much of going to England, feel quite embarrassed to decide. Feel however on the whole inclined to stay and run my chance in America.

February 18. Still in doubt about removal to Europe. My plan for this country is to save sufficient for a year's rent of a store, put all our furniture in it, and then hope and try for the best.

February 19. Determined to write and explain my views to my father, and then decide according to the tenor of his answer.

March 11. Mrs. Sigourney's 'Letters to Young Ladies' a well-written and instructive work, indited in the right spirit and likely to effect much good.

March 28. Trade very dull, with but little hope of amendment.

May 8. Which would be better, to go to England, or become a teacher ?

June 18. Was asked to-day if I would avow

my scepticism as regards the duration of matter or the eternity of a creator. Answered affirmatively, which I now think was foolish as I feel inclined to lend more faith to the Bible.

December 22. Feel more satisfied as regards our circumstances and begin to think we ought to hope for the best.

1839.

January 17. Poetical idea returning, encouragement a stimulus to further exertion and improvement.

January 24. Feel that I have been too lazy and too extravagant of late. By aiming at too many objects one risks the chance of losing all. Steady application required.

January 28. Still very cold, strong wind from N.W. Requires some courage to work out of doors. To Library au soir and took ' Pursuit of Knowledge Under Difficulties.'

January 29. Settlement of my yearly account, does not appear as satisfactory as could be wished, but perhaps after what has passed we may esteem ourselves fortunate.

January 30. Reading, writing, working at wash-stand au soir. Much pleased with the book I have showing that knowledge may be acquired in spite of adverse circumstances. Ought I not to apply myself also ?

January 31. Feel rather unsettled with literature and business. Why not apply myself with energy to the latter?

February 1. Poetical ideas present themselves. May I cultivate them to good purpose with satisfaction and profit.

February 10. Reading 'Memoirs of Knebel,' specimen of early German comic romances, 'Robbers' of Schiller's. Must it be inherent or can it be acquired, a keen power of perception and discrimination ?

March 30. Letter from Reading containing explicit details and inviting our immediate return. Dieu dispose de tout.

May 12. Felt grieved at leaving behind us the bones of an offspring.

His daughter, a young child, had died during their residence in America.

May 20. Sailed for England.

June 12. Entered the docks, and landed at 4 p.m.

June 17. Started for Reading . . . five years' absence has not diminished our affection.

He settled down there once more in his father's business, but work did not readily come, and until the end of the year are entries " Trade dull." From this time the diary breaks off until January 1841, but the pages are utilized in the intervening year as a commonplace book of passages that had struck him in the works of others. Thus we find in August 1840, from ' Hints to Mechanics '—

And as to *luck*, as I have said before, there is more in the sound of a word which people have got used to, than in the thing they are thinking of. Some luck there is no doubt as we commonly understand the term, but very much less than most persons suppose. There is a great deal which passes for luck which is not such. Generally speaking, your *lucky "fellows,"* when one searches into their history, turn out to be your fellows that know what they are doing and how to do it *in the right way.* Their luck comes to them because they work for it; it is luck well earned, they put themselves in the way of luck. They keep themselves wide awake. They make the best of what opportunities they possess and always stand ready for more; and when a mechanic does thus much, depend upon it, it must be *hard* luck indeed if he do not get, at least, employers, customers, and friends.

" They need only," says an American writer, " to turn to the lives of men of mechanical genius to see how by taking advantage of little things

and facts which no one has observed, or which
every one had thought unworthy of regard, they
have established new important principles in the
acts, and built up for themselves manufactories
for the practice of their newly discovered pro-
cesses." And yet these are the men who are
called the *lucky* fellows and sometimes envied as
such. Who can deny that this luck is well
earned ? or that it is just as much in *my* power
to " go a head " (as the Yankees say) as it was
in theirs ? Knowledge, acquisition, effects of
Education of the Senses.
Languages, the importance of.

Again, in September 1840, he notes from
' Mental Culture '—

To commune with great minds through their
works is the secret way to become rich and
powerful in thought, and to have expansive and
noble views of truth. Who would talk with a
novelist, when he might hold converse with a
Milton or a Locke. In reading the thoughts of
such men thoughts greater than the growth of
our own minds are transplanted into them and
feelings more profound, sublime or comprehensive
are insinuated amidst our ordinary train, while
in the eloquence with which they are clothed

we learn a new language worthy of the new ideas that are created in us.

Of how much pure and exalted enjoyment is he ignorant, who never entertained as angels, the bright and lofty emanations of loftier intellects than his own ! By habitual communion with superior spirits, we are not only enabled to think their thoughts, speak their dialects, feel their emotions, but our own thoughts are refined, our scanty language is enriched, our common feelings are elevated, and though we never attain their standard, yet, by keeping company with them, we shall rise above our own as trees growing in the society of a forest are said to draw each other up into shapely and stately proportions, while field and hedgerow straggles, exposed to all weathers, never reach their full standard, luxuriance and beauty.

Intellectual culture consists not chiefly, as many are accustomed to think, in accumulating information, though this is important, but in building up a force of thought which may be turned at will on any subject on which we are called to pass judgment. It is not so important to know everything, as to know the exact value of everything : to appropriate what we learn, to arrange what we know.

The same month he makes a note from one of Jeffreys' writings—

Simple pleasure is the best, when the inordinate hopes of youth, which provoke their own disappointment, have been sobered down by longer experience and more extended views; when the keen contentions and eager rivalries which employed our riper years have expired or been abandoned ; when we have seen year after year the objects of our fiercest hostility and of our fondest affections lie down together in the hallowed peace of the grave; when ordinary pleasures and improvements begin to be insipid, and the gay decision which seasoned them, to appear flat and importunate ; when we reflect how often we have mourned and been comforted, what opposite opinions we have successively maintained and abandoned, to what inconsistent habits we have gradually been formed, and how frequently the objects of our pride have proved the sources of our shame, we are naturally led to recur to the days of our childhood and to retrace the whole of our career and that of our contemporaries with feelings of far greater humility and indulgence than those by which it had been accompanied ; to think all vain but affection and honour, the simplest and cheapest pleasures the

truest and most precious, and generosity of sentiment the only meat of superiority which ought
either to be wished for or admitted.

From Sharp's letters and essays he quotes the
following passage—

There are few difficulties that hold out against
real attacks, they fly like the visible horizon
before those who advance. A passionate desire
and unwearied will can perform impossibilities,
or what seem to be such to the cold and feeble.
If we do but go on, some unseen path will open
upon the hills. We must not allow ourselves to
be discouraged by the apparent disproportion between the result of single effects and the magni-
tude of the obstacles to be encountered. Nothing
good or great is to be obtained without courage
and industry, but courage and industry might
have sunk in despair, and the world must have
remained unornamented and unimproved if men
had nicely compared the single stroke of the
chisel with the pyramid to be raised, or of a
single impression of the spade with the mountain
to be levelled. All exertion too, is in itself delightful, and active amusements seldom tire us.

The greatest question is, what will others
think ? not what is in itself truly good or beauti-

ful, or in itself despicable, debasing or pestilential. This is a state most unfavourable to worth of character, there can be none of the higher virtues, there can be none of the finer qualities in, and formed under such influence. There must be a rising above these considerations to achieve anything which shall command the homage of our hearts and which shall beam on our minds like a reflection from the moral image of God. " What will Mrs. Grundy say ? " is often a far more potent question than, " What is the voice of God speaking in man's own soul and conscience ? "

Reading without intelligence injures the brain and stomach mechanically. Reading with intelligence injures both in the less direct way of nervous excitement ; but either way much reading and robust health are incompatible. Only let a child eager for knowledge be read to instead of allowing him to read himself, and the whole of the mechanical mischief is avoided : and again let him be freely conversed with in desultory manner in the midst of active engagements, and out of doors, and thus, while an equal amount of information is conveyed, and in a form more readily assimilated by the mind, nearly all the mischiefs of excitement as springing from study,

are also avoided.   In a word, let books in the
hands, except as playthings, be as much as
possible held back during the period of early
education.

> " I would not live in fear—
> I would not hold existence on the bond,
> That like a coward, I must lie for life.
> This for myself ; but for mankind at large,
> I leave them where I found them—it may be
> With errors of some service, in a state
> So full of errors—nor would teach a truth
> Might work like falsehood on perverted minds.
> The toiling world in uneasy movement wait
> Beyond all skill of mine to tamper with—
> Moves or revolves, as God ordains.   My task
> Is with my single heart, and in own truth,
> I cannot struggle with mankind in arms,
> Nor find out truth for all."—ATHELWOLD.

Early in 1841 a few paragraphs of diary are
found again.

1841.

January 25. Do not feel assured of success in remaining here; so many counter circumstances.

February 11. Letters from New York— went to London about maple, saw Mr. Eales, enquired about New Zealand. Touch of pleurisy, not very favourable prospect of business here.

February 26. Convalescent, been working the past twelve days. Resolved to see nothing for one year but exertion and faire de l'argent, so that if we go away we may not have to reproach ourselves for idleness. Seasons in New Zealand —Spring commences middle August. Summer, December. Autumn, March. Winter, July. The island contains 95,000 square miles, or 60,000,000 of acres.

April 16. After much anxiety, I hope the maple business will be settled. . . . too much selfishness in my nature which I must endeavour to eradicate.

D

November 24. Must think of emigrating, no prospect of amelioration here.

December 10. Suicide of Religious Freedom Society is that genuine Christianity that will make ministers of the gospel insult another on a public platform. Wrote a few remarks on the subject, and sent them to Underwood for revision. Still here; my Lecture on Cabinet making [at the Reading Mechanics' Institution] very favourably spoken of, which does not elevate me in my own estimation as I so well feel my own ignorance. I do not know half so much as I ought, neither do I turn what I *do* know to the best account. How much may be accomplished by energetic exertion and assiduity. The idea of a lecture on " Knowledge and Information " has occurred to me—if well developed it would be a useful and highly interesting subject. General depression in trade, many failures.

1842.

February 5. At beginning of last month paid a visit to London, did not succeed in seeing the maple . . . in correspondence with the Venetian Candle Company respecting an agency. Heard Lord Chelsea try to speak at the meeting for the Athenæum.

May 22. Last Wednesday finished second month of singing . . . business very dull indeed, just creep along in the workshop. Much trouble to find a house, being desirous of removing.

October 10. Heard from Cowdery that I am to go to work with him in London next Monday. No hope of doing anything here, so that the other will perhaps be an opening to something better. Must work hard to leave all in order by that time.

1843.

July 20. Since the above was written have been in London as clerk to Mr. Mainzer, author of 'Singing for the Million'; like my employment very much, but the remuneration is not sufficient for me to keep my family in town. I have been to Reading to see them three times since last October, hope to go again in a week or two. Have had many letters from New York, but do not succeed in selling, business so very dull. Have recently heard two very excellent sermons from Davey, esteem myself fortunate to have had the opportunity of listening to his words.

September 24. Have been reading 'Self-formation' for the second time, and with increased pleasure. The perusal of its pages has removed many doubts and difficulties from my mind, and given me a degree of courage and confidence for the future. I have had occasional fits of gloom and despondency lately, the causes of which I have endeavoured to trace, and I think their

operation is pretty nearly as follows. I have a
constant desire, a restless craving to make my
pen obey all the thoughts of my brain, nothing
but the realisation of this object will appease
this unquiet feeling. Why I cannot do so now
is, I suppose, because I do not *think* correctly.
Because my thoughts are only half formed, and
do not present themselves in a shape which may
be transferred to paper, and what is ill-conceived
by the brain must be badly executed by the
hand. Oh, this continual straining of the mind
after higher and higher achievements! from
what does it arise? From love of fame, from
the wish to make a name, from the pressure of
poverty, or from a real love of literature? I
think I can set down the latter as the *primum
mobile*, but will take time to reflect. I sit down
to versify, I find my ideas flow poetically,
pleasantly and freely; when suddenly my mind
becomes a blank, a worse than blank, a very
chaos of confusion, and the writing which before
was a spontaneous effort becomes mechanical
drudgery. This is a great cause of despondency.
I can scarcely help thinking that my mind must
be deficient in some very essential quality, which
will ever prevent the half-formed creations of my
mind from becoming visible realities, though I

have reason to believe that in this particular
I have some ground of hope, for I have read
often that " Industry is the best genius." . . .
The day when occupied in business, the privation
is little thought of, but the evening awakens a
longing for home, its joys and sorrows, its
sympathies and endearments. The absence of
this grateful and prized repose or termination to
the day has doubtless a depressing effect on the
mind, which is now kept in a state of tension
or dullness unrelieved by the genial influences of
home; the voices of my children, and approving
glances of a wife.

Another cause of gloom I think arises from
the feeling that grows upon me every day,
that I know nothing. I used to think that
I was possessed of a good share of information on most subjects, but I have found that
information is not *knowledge*. My residence in
London has made me sensible of my ignorance.
I seem hitherto to have lived in vain. Have I
made the best of my opportunities? Had I
done so should I be the dull unoriginal being
I am? Hence the reaction of gloom, hence the
impression of doubt aggravated at times by
the hopelessness and misery of despair. Yet
have I *real* cause for despair? I am here in the

possession of health and strength, will not the
capacity for thinking increase and widen by
cultivation and exercise? If the phrases "Prero-
gative of genius," and "Predominance of the
natal star" be, as I have read, but "Cabalistic
nonsense," have I not then cause for hope,
cause to confide in the promise that "diligence
overcomes all"? In 'Self-formation' I find that
reading is of little use, unless we endeavour to
turn it to account as we go on by agitating it,
by meditating on it, by casting it into different
moulds and viewing it under every variety of
form. Composition, whether rhymed or in
prose, must be one of the means by which we
test the progress of our mind, by which we may
attain clearness of conception, and give to vague-
ness a vivid and definite significance.

The exercise of the faculty of composition must
be attended with ease, however seldom it may
be practised. If by pursuing it I can achieve
the mastery over my imagination, and render it
subservient to my will, verily I shall not be
without my reward. To trace the operations of
the mind must be a great aid to composition.
I will endeavour to trace the history of mine
from the time of my earliest recollections to
the present. I will endeavour to elucidate and

illustrate the causes of its feebleness, incompre-
hensiveness, and indecision, or at least to show
that it might have become more efficient than it
is. To do this I shall occasionally write a page
or more as the time and feeling of the hour may
serve. This will not interfere with any of my
present pursuits, as it can be done to fill up odd
minutes of time, fragments of a day which
would otherwise be lost. But here again the
thought comes, will this be doing justice to your
family? might you not be better employed in
seeking out means to increase your income? so
that the bitter necessity of living separated may
be dispensed with. But on that subject I have
thought myself dry, not a single new available
idea have I been able to raise after months
of severe cogitation; there, I do not think I
shall be guilty of any dereliction of duty in
examining the history of my life and heart. If
this be done in the proper spirit it must qualify
me to become a better citizen, husband, and
father.

I have now been writing for an hour, with
occasional stoppages, in which I have laid down
my pen, and leaning back in my chair have
travelled away into a whole wilderness of
thought. I did not expect when I began that I

should write more than half a page, but my
spirit seems to have taken to the work kindly;
my pen has gladly obeyed its dictates, and lo,
four pages are blackened, whether for good or evil
the result will show. At all events I must take
this preliminary essay as a favourable augury for
the future. Why is it always easier to describe
what we have felt than what we feel, or why, as
the French express it, " On rend mieux conte de
ce que l'on a senti, de ce que l'on sent " ? I view
it as almost an easy task, this of retrospection
which I am about to undertake, but if I were to
try to describe some present feeling or circum-
stance I should be at a loss for the requisite
ideas. Is it that we require the comparison and
experience of the present to enable us to describe
the past, and thus facilitate the performance, or
is it that this indulgence in the past is a labour
of love to memory which imperceptibly guides,
influences, and quickens its present operations ?

Walter White here gives his youthful recollections quoted at pages 1—5.

November 11. Received at noon to-day notice
to start for Scotland per mail train this evening,
consequently there was a sudden and hasty
packing and clearing out and sending things to

Kennington for safe keeping. Left London at a quarter to nine. Reached Birmingham at half-past one. Thence at two by first-class to Liverpool, where arrived at six. Dived into a cellar till daylight, then, after breakfast, a good wash, etc., removed some of my drowsiness, made my way to Clarence dock and engaged a berth in the *Admiral* steamer for Glasgow. Wrote to Chambers, out at a quarter after one, weather so thick could scarcely see any thing. Sea quite calm at nine. Same evening saw lights of Douglas, Ramsay, and Point of Ayr, then turned in till daylight, when we were in the estuary of the Clyde off Rothesay. Reached Greenock at seven a.m., thence by small steamer to Glasgow. Scenery on the Renfrew shore bold and fine. The railway runs along close to the water. Passed Port Glasgow, and the castle, a very small building in a low situation; in this the unfortunate Mary was some time a prisoner. The channel of the river is here very narrow in comparison to the whole width of the stream, and the massive stone wall, surmounted with towers, having a cross on each summit, has while it shows the course of the stream, a very picturesque effect. The embouchure of

the Leven, crowned as it is by the bold rock
and ruined fortress of Dumbarton, is very
beautiful, but I looked in vain for the "lofty
Ben Lomond," the atmosphere was too thick,
and completely hid the mountain monarch;
however I was shown a place in the clouds
where I was assured he would be visible if the
weather were clear, and with this I was obliged
to content myself. The view of the rock greatly
enhanced Wallace's daring and successful exploit
in my estimation. None but men fighting for
their liberties could have achieved such an
enterprise. Not far from Dumbarton on the
same shore is a monument to —— Bell, but I
did not learn which individual of that name
was thus honoured. Also the entrance to the
Forth and Clyde Canal, about sixty miles long.

Various fine private mansions now began to
indicate our approach towards a grand centre of
commerce and civilization, and we passed on our
right the mouth of the Cait, a small stream that
flows through Paisley, which lies about four
miles from the Clyde. Here we caught sight
of the church and more prominent buildings
of Renfrew. The river now becomes very narrow,
much more so than I had anticipated. I had
always imagined the Clyde to be a noble stream

at Glasgow. After a voyage of two and a half hours from Greenock we arrived at the Broomielaw, a very fine quay by the side of which were moored many large and splendid steamers. Now began the hurry and confusion of searching for baggage, while the uncouth and eager cries of the crowd of boys and porters on the wharf made a diverting scene. No sooner were we along side than it appeared as though each passenger must resign all claim to his own personal effects, so industrious were the blue-bonnets in appropriating them, while the "noddy" drivers on their part were equally noisy and effective in making their claims known. However I fought my way through, though at one time not less than a dozen boys were pulling at my carpet bag, but as competition was so great I found no difficulty in getting a boy to carry my bag a mile for a penny.

On leaving the steamer there was the usual hasty farewells of transient acquaintances, and I could not help thinking how people are thrown together, converse, give and receive experience, reform and renew ideas, and sometimes lay the foundation of a lasting friendship, and then are separated, never to see each other again. Here I had met with a gentleman from Wolver-

hampton with whom I talked over my recollections of persons and places I had known
when there in 1832. I found him intelligent
and communicative, but disposed to be sea-sick,
out of which I roused him by continued walking
and conversation. He remarked that the exercise
was beneficial, as it was the first time he had ever
been to sea without experiencing its sickening
effects. But to return. I went to the Bath
Hotel, where I delivered to the landlord, a
pleasant fellow, the communications with which
I was charged, and then sent for Burns to make
enquiries of him as to the state of affairs.
Under his guidance I walked about the town,
saw the Cathedral of St. Mungo, which stands in
much need of repairs, the Necropolis, Bridge
of Sighs, and the public buildings, and the tall
chimney 478 feet in height. After dinner the
landlord came in to chat a little, and we drank
a glass of toddy together. At five p.m. I started
for Edinburgh, 44 miles, passed Kirkintilloch,
Falkirk, Linlithgow. A most bitter ride, the
country being very open and exposed.

Reached Edinburgh at half-past seven, and was
glad to walk to St. Andrew's Square, as that
restored a little life to my limbs. Found Mr.
Mainzer busy, the whole table covered with books

and manuscripts, and soon was engaged in helping him, with occasional breaks in giving and
receiving intelligence.   Was not sorry to pass a
quiet night in a comfortable bed, and rose in the
morning all myself again, and found enough
occupation in making the arrangements for
his lecture in the Music Hall, George Street, at
two o'clock, which, although quite a stranger, I
succeeded in getting completed in good time.
This duty brought to my mind the memory of
the days when I was perambulating the country
a Hamiltonian teacher, though not so successful
in a pecuniary point of view as could have been
wished, considering the profits were for the funds
of the fever hospital.   Yet the lecture was well
received, and excited a great deal of interest.
The next day entered on my duties at the
Music Hall, St. David Street, and dismissed the
former attendant, who at two o'clock in the day
is most offensively redolent of whisky.   In all
this time ideas of my position have been flowing
in upon me, and I hope I shall understand my
duty and shall be able to perform it.

On Wednesday morning it was what they call
here a "saft day," anglice, raining like fury.
Walked out to the residence of Dr. Chalmers
at Morningside, who received me very politely

and gave attention to my communication, and promised to lend his assistance to the cause for the improvement of Psalmody. In this walk although the weather was so bad I was delighted with the fine bold prospect I had of Salisbury Crags, Arthur's Seat, and the Pentlands. In the evening of this day I called on Adam Black, the new Lord Provost, upon Sir James Forrest, a fine affable old gentleman, and J. Learmaith, Esq., to confer with all respecting the lecture to be given on Friday evening. Thursday and Friday I was busy in calling upon clergymen, on the evening of this latter the second lecture was delivered in the Music Hall, quite a failure in a pecuniary point of view, not two hundred paying auditors, and about three hundred free, and although some families came provided with several free tickets, yet they wished to pass other members of their families free also, a stretch of meanness which I resolutely refused to submit to; pay or stay out was my rule without exception. I stepped upstairs towards the conclusion of the lecture and heard M. sing two ancient melodies. One, " Eine feste Burg ist unser Gott," by Luther, was sung by all the Protestant armies of Germany during the Thirty Years' War. Sung by M.'s expressive voice they were admirable, and were duly appreciated by the audience. After the lecture there were speeches by Lord Murray, Mr. Simpson, and others, and letters were read all expressing their concurrence in the efforts making for the improvement of Psalmody. Saturday was a meeting of the Students of Divinity in the same hall. I met Mrs. M. who arrived by coach, attended committee meeting at four, and was busy until quite late in folding and dispatching circulars to clergymen. In consequence of Mrs. M.'s arrival, had to seek a lodging which I found at 15 St. James' Square, two rooms comfortably furnished, including fire, attendance, and boots, 7/6 per week. Took possession at six p.m., and found myself quite at my ease, but could not help contrasting my solitary state with what it would be were all my boys round me.

November 19. Went to hear Guthrie preach. Not so eloquent as my anticipations led me to expect.

Sunday 26. Walked by accident into the Gaelic Church and heard a most eloquent sermon in that language, of which unfortunately I did not understand a single word. The manners of the congregation very beastly.

From this date until February, White was occupied in Edinburgh, visiting libraries, hearing sermons and lectures—amongst the latter Professor Wilson—and in writing circulars "and otherwise preparations" for Mr. Mainzer's candidacy for the Chair of Music now vacant by the resignation of Sir H. Bishop.

1844.

On February 12 he writes—

From December 22, 1843, until this date from various reasons have made no entry; consequently I must now write down recollections of facts and things and give simply a summary. My time has principally been occupied in translating the 'Sketch' of Mainzer's life from the French, in which I succeed better than I anticipated, and would like to try my skill on some more important work. Have also had to translate and arrange French letters sent as testimonials of which two publications have issued. Read various books, Cooper's latest works, the magazines, Thierry's 'Histoire de la Conquête,' a most interesting work, one which gives a greater insight into the early history of Britain than any other I have read. What light he throws upon all the circumstances relating to the conquest; he draws also a very different picture of Alfred to the one usually given. . . . By the kindness of *Millen,* one of the clerks in Oliver and Bond's house, was shown over their establishment, which was seven or eight acres. A portion of it comprises a house which formerly belonged to the Marquis of Tweeddale, the painted doors to imitate books are still to be seen in what was the Library. The establishment is not so imposing from appearance as from its great extent, and the multiplicity of operations carried on within it. I saw bookbinding in all its stages occupying a great many hands and all very busy, also machines for cutting boards and rolling books, and hydraulic presses, the largest of which exerts a pressure equal to a thousand tons. There are three steam printing presses, two of them of a construction which I never saw before, and at which two works can be printed at once without any loss of time or power. The pressure works up and down steadily in the centre, while on each side alternately the machinery on which the sheet is laid, works in and out with the most beautiful precision. . . . I was introduced to one of the firm. Professor Wilson is a fine-looking old fellow, he is slightly lame and walks with a stick but appears vigorous for his years; he has an imposing appearance, being tall and stout; wears a broad-brimmed hat from under

which his long flowing locks appear and give him
almost a patriarchal look.   He was elected Professor in 1820.

I have attended Mr. Simpson's lectures, the
course of which is now extended to twelve
nights instead of eight.   The attendance is
good but not so numerous as at first.   It is
however very creditable to the working classes
of Edinburgh that three thousand of them came
forward and signed a requisition for eight lectures
on their moral and social improvement.   Mr.
Simpson, who is a great friend to all such movements, gladly acceded to the request and has the
satisfaction of knowing that in many cases his
advice has been followed.   He lent me a book,
written by John and Alexander Bethume, two
peasants, on practical economy, it is a work that
would do credit to men of greater opportunities
and more eminent station.

February 13. Interesting lecture from Mr.
Simpson in the evening.   After a few remarks
on " Love of Approbation " he proceeded to
explain and illustrate the superior qualities, the
moral feelings, those which comprise the whole
circle of ethics, viz. benevolence, conscientious-
ness, veneration, which are so admirably expressed in Scripture in the text " Do justly, love

mercy, and walk humbly with God." These
are, as Dr. Chalmers said, the seal and the impression, they mutually confirm each other, the
religion, and the philosophy. An anecdote of
Massillon was adduced as an instance of veneration. One day while he was preaching, a
tremendous thunderstorm came on. He essayed
to speak, when a crashing peal overpowered his
voice, again and again with the same result.
He then said, " When the Master speaketh, it
becometh the servant to be silent," and sat
down; such was the feeling of veneration induced by his impressive words, that he did not
require to preach more on that day. Some excellent remarks were made concerning *apparent*
sanctity, the regular habit of churchgoing, etc.
as sufficient for all the requirements of Christianity. What is not benevolent is cruel, what
is not conscientious is dishonest or unjust; what
is not veneration is impious.

February 14. Called on Hugh Miller with a
letter of introduction. He wrote those articles
in 'Chambers' Journal,' the "Gropings of a
Working Man in Geology" and "The Old Red
Sandstone." Though it was 11 a.m. when I
called he was not out of bed, as I learnt that
he is now editor of the 'Witness,' and last night

being publishing night, he lies later in consequence this morning. However, I presented my credentials to his wife, and after chatting with her a few minutes, he made his appearance, unshaven and rough-headed. He is a tall, powerful man. I could not help thinking of the days when he was quarrying at Cromartie. The fact of his being now an editor seemed to diminish the interest; possibly because of the incongruity between geology and the stormy columns of a newspaper. However, he invited me to inspect his collection, which is ranged round the walls of a room which seems to be at the same time his library, and he kindly gave me explanations of his specimens, their order and arrangement.

February 15. By the kindness of Mr. Simpson, who permitted me to use his name, I obtained a sight of the establishment of Messrs. Chambers. On presenting myself at the door and enquiring for Mr. Dickson the foreman, I was requested to wait a few minutes, when he made his appearance and very kindly and politely expressed his willingness to conduct me over the premises. From the apartment in which I was waiting and which is entered from the close, by a door generally kept locked, he led me to the floor

below, the case room; here several men were at
work, and I remarked that they had a much
cleaner appearance than is generally seen in
printing offices. The sale of the various works
published by the firm is such as to keep all
these men in constant employment. Here the
foreman proceeded to explain to me the various
details of a printing office, with which I informed
him that I was already acquainted, and that in
looking over an establishment like the one in
question, I felt that I regard the *morale* more
than the practical, or rather, that the gratification
induced by the sight of the latter was heightened
by the reflections suggested by the former. In
passing from this room I noticed the large school
maps stretched on frames and lined, undergoing
the process of drying. The frames in which they
slide have the advantage of drying a great number
at once in a small space.

At this point we may pause to consider Walter
White's position. His life had been a prolonged
struggle not only to fit himself for a higher position than that of a skilled workman, but incident-
ally to earn a livelihood for himself and his
family. His self-education had been marvellous;
and the better it grew the less tolerant he
became of his life as a cabinet-maker. He
was about to find a position which was better

suited to his needs and aspirations; and as his diary shows he took full advantage of his opportunity. The next entry in his diary indicates the direction in which his talents were to turn; and from this point his journal assumes the character which justifies its publication. His work now lay among the *savants* of his time. He notes their peculiarities and their differences; he records their sayings and doings when they call at the rooms of the Royal Society. But with all this, he illustrates a very curious contrast in his own life—the contrast shown by his life as a workman and by his life as a servant of the Royal Society and ultimately a friend of many of its Fellows. In effect his diary tells its own story. In April 1844 he was thirty-three years of age.

It is interesting to read the letter which procured Walter White his appointment at the office of the Royal Society. The letter is written by James Simpson, Advocate of Edinburgh, to Mr. Weld, the then Assistant-Secretary of the Royal Society, and is dated Edinburgh, Sunday, March 18, 1844. It reads—

"My dear Sir, I have taken a great interest in the bearer, Walter White, a Reading (Berkshire) man who has been here all the winter in an employment where I have seen a great deal of him. I was particularly struck with the quiet efficient activity and discretion of all he did—and with his power (a rare one) of following a matter out, when the object is made known to him, and with his general intelligence and information. He is a self-educated man, but better educated than many gentlemen. He is a perfect French scholar, speaking the language like a Frenchman; and he knows something of German and Italian; he has some Latin. He understands books and can manage a library, and would be a most valuable *employé* in many situations. . . . It occurred to me to put Mr. White in your way. He wants bread for himself and family no doubt; but to whomsoever he is indebted to for it, he will make an ample return in valuable services. His employer here speaks of his moral character in the highest terms. He is well worth your looking after, even if nothing occurs to you that he might do at present.

"P.S.—Mr. White wrote an excellent poem on "Channing" in 'Tait's Magazine,' December 1842, p. 778. He attended my lectures this winter and will tell you all about them."

This excellent testimonial and singular sketch of a man's character was enclosed in a letter from Mr. E. Chadwick to Mr. Weld, dated April 3, 1844. It runs thus—

"My dear Sir, I send you a person to look at.

The enclosed is his recommendation from Mr. Simpson of Edinburgh, Advocate. Mr. White was brought up as a cabinet-maker and has acted as secretary to Mr. Mainzer, the lecturer in music."

To these letters Walter White owed his eventual appointment, first as Sub-Librarian of the Royal Society and subsequently as Assistant-Secretary on the retirement of Mr. Weld. The period at which this change in his life occurred, marks the development of the working man into the man of letters. His success was mainly due to the characteristics pointed out in Mr. James Simpson's letter. Henceforth his life ceased to be a long and unhappy struggle for existence. In London he was soon to occupy a place at the very centre of the scientific and literary world. His diary faithfully reflects the change, and illustrates more completely than any comment, the contrast between the working man and the man of letters.

CHAPTER II

SUB-LIBRARIAN AT THE ROYAL SOCIETY, 1844—1853

1844.

April 19. Entered the service of the Royal Society.

May 19. Have now been one month in my situation as sub-librarian at Somerset House, should like the occupation better if it were more intellectual, but hope to render the advantages afforded by my employment subservient to my intellectual progress. Now that I am placed in a position which I have long desired I must not go to sleep, but make it the basis of more strenuous and successful efforts. The duties required of me are not onerous, and I am well contented with the Resident Secretary, Mr. Weld. I look forward with pleasure to the reunion with my family at Michaelmas, and in this hope we may yet bear a few months of separation. Have I not reason to be happy and thankful, what a load is taken from my mind? I am now humanly speaking in a permanent situation; on waking in the morning I have no longer the harassing dread and uncertainty as to the future, my thoughts are free to dwell on whatever bright ideas may present themselves, and happy am I in the contemplation of the society of my family. May I ever retain a just sense of the benefits I enjoy, and strive to render myself more and more worthy of them.

October 7. Have received several letters from my very kind friend Mr. Simpson, whose interest in my welfare continues unabated. He has sent two of my papers to Messrs. Chambers, which have been approved; and will doubtless be published. This is encouragement for me to go on._ I am also earnest to pursue the study of German, in which I have made some progress. ... Am working at a series of papers on America; reading also the 'Health of Towns Report' presented to me by Mr. Chadwick. Must not forget to say that I feel happier now than I have felt for years.

Subjects of papers sent to Messrs. Chambers:

'A Library: Old Books.' } Approved.
'Go ahead.'

'Five years in America.' (Now in progress.)

'Keeping down.'
'Don't tell Father.' } To be written.
'The Pawnbroker.'

Monday 28. Saw the Queen go in state to open the New Exchange, rather a sorry affair. Have read a work called 'Vestiges of the Natural History of Creation' in which I am much interested. Saw the Earl of Rosse on Saturday, he is a plain well-grown man, farmer-like in appearance.

November 24. Have been informed that the Queen looked very humble on the occasion of opening the Exchange, she seemed pleased though oppressed with the extreme enthusiasm with which she was greeted, possibly she may have felt that too much worship was paid her. She had, it was remarked, a troubled look in her beautiful eyes.

1845.

January 1. The past year has ended, and left me happier and more hopeful than for a long time. I have made more progress, and feel myself more capable of doing so in future.

January 2. Walked to Bishopsgate Street with veneers for Mr. Snowdon. Very fine day, but feel rather dull and ill-tempered.

January 3 *and* 4. Letter from Mr. R. Chambers signifying his approval of my papers on 'Matteucci' and on 'Health of Towns.' Have made pretty good use of my holiday. . . . Shall go to office again on Monday all the better.

January 10. This day I was agreeably surprised on receiving a remittance of £5 16s. from Messrs. Chambers. They reward very generously my humble attempts, and encourage me to go on.

January 25. A very encouraging letter from Mr. R. Chambers. What do I not owe to Mr. Simpson's kindness! All this must stimulate me to make a good use of my faculties, to correct my heart, and to perform my duties well in

every station of life. I know that these will be
not mere empty words, but the expression of an
active and energetic principle. Can we deceive
ourselves? Do I not feel that I am not what
I ought to be? There is great room for
improvement. Mr. Simpson has written to con-
gratulate me, and Mr. Chadwick has called to
say he will do all in his power to second Mr.
S.'s views. Give me a live and grateful heart,
Oh! Thou.

January 31. Heard Faraday lecture at the
Royal Institution on the Liquefaction and Solid-
ification of substances commonly considered
gaseous. Saw for the first time in my life
Carbonic Acid in a solid state; the method
by which he produces an inconceivable degree
of cold was beautifully explained and illustrated.

March 1. Have now finished the search for
the titles of old books in the Library Catalogue;
to me it was a task of much improvement and
has given me much insight into the history of
books, as well as an acquaintance with the names
of their authors. How agreeably time passes
when we are engaged in occupation congenial
to our feelings. Life has now for me a certain
object, I hope by my humble attempts at author-
ship to be of some use in the world and to

perform all my duties faithfully. I fear I have been too peevish and overbearing lately; how necessary it is to keep a constant watch over one's temper, thoughts, and heart. Have another encouraging letter from Messrs. Chambers specifying other subjects on which I may try my hand. Truly the kindness of these gentlemen is refreshing and gladdening to the heart; may I ever deserve it. Saw to-day Sir H. Pottinger. Sir John Franklin is to go out on the new Arctic Expedition in May. He seemed to be a fine kind-hearted old gentleman, and though advanced in life, must be of an active disposition to be willing to brave once more the dangers of the Polar regions where he has already suffered so much.

March 20. Have been very busy to-day preparing the magnetic observations for dispatch to the Foreign Observatories, and hope Sir John Lubbock will have no reason to complain of slowness.

October 10. Speculation in railways is now as rife and promises to be as fatal as in the South Sea Scheme. A new project, the "National Railway," capital £40,000,000, is said to be all subscribed for. Steamboats now ply from London Bridge to Westminster for a penny.

October 15. Saw Henry Stone* on Monday. He told me that Miss Mitford knows Wordsworth, and finds the littlenesses and meannesses of the man completely spoil her veneration for the poet. One of his daughters married a wine merchant, and wherever he visits it is expected that the parties honoured by his presence should buy their wine of his son-in-law.

October 16. This year was published the anonymous work entitled 'Vestiges of the Natural History of Creation.' No book since 'Waverley' has caused so much surprise or discussion. Whoever the author may be he has made not a few remarkably shrewd guesses which have created not a little excitement in the scientific world. Nearly the whole of the first edition was given away, many of the leading men of the Corn Law League were presented with a copy. The fourth edition is now published.

The new hall in Lincoln's Inn Fields is to be opened on the 30th. It is said that the wives of the Benchers are such "low creatures" that not one lady will be present to be introduced to Her Majesty, who will preside.

Dr. Miller elected this year to the Professorship

* A Banbury bookseller of superior intelligence, and a lifelong friend of Walter White.

F

of Chemistry in King's College, in the room of Mr. Daniell, deceased.

October 25. Hear that Dr. Miller has not more than two pupils, while Mr. Graham of University College has a great increase. What numerous newspapers have started into being under railway excitement. Men with models of locomotives on their heads parade the streets to draw attention to their placards.

November 22. On Thursday was our first meeting for the session, three gentlemen were blackballed. Faraday's paper was commenced on light and magnetism, a very crowded meeting. The committee which awarded the royal medal to Mr. Beck are to reconsider the grounds of their decision.

November 29. Walked with Harry and Hamilton [his sons] to Charing Cross, thence by omnibus to Great Western Rail, where I sent the two little fellows to Reading. Grieved was I to part with them, the trial was sore indeed. Nothing short of a sincere conviction of its necessity could have supported me.

Dr. Buckland, the new Dean of Westminster, was one day speaking on some geological subject to the Geological Society when, after stating his views, he said, "If you don't believe it, may you

itch and never be able to scratch yourselves."
It is said the Oxford professors rejoice at his
promotion, saying, "Thank God he is gone, if
we could only get rid of Daubeny we should
not have a scientific man amongst us."

December 11. Yesterday the whole of the
Ministry resigned, and report says Lord John
Russell has been sent for. It has taken every
one by surprise, the obstinacy of the Duke of
Wellington was the cause. Cobden must form
one of the new Ministry if it is to be permanent.
On Wednesday H. Stone came up. . . . He
gave me a hint as to an article on Proper Names
in Poetry.

December 22. Took leave of our snug little
house in which I had hoped to pass many happy
days. Has our separation been occasioned by
my faults? Could I have prevented it by keeping
a better watch over my words and actions?
There was much cause for reflection in taking
the last look of the dismantled dwelling, and
walking for the last time up the Old Kent Road.
We grow insensibly attached to places and
objects, and at last their loss is a cause of pain
to us.

1846.

January 26. Saw the first number of the 'Daily News' yesterday. Like its tone and their expression of a desire to elevate the character of the public press of this country. It is said to have a fund in reserve of £180,000. Charles Buller is editor, supported by Talfourd, Forster, Dickens.

February 18. Bulwer is said to be the author of the 'Modern Timon,' and H. F. Chorley about to bring out a tragedy that will stamp him as *the* dramatic genius of the age. Lord Brougham keeps a man to make extracts from books, to whom he gives his orders thus, "Gut that book for me," from which it may be said that his lordship's writings issue from a literary gutter. Heard R. Brown tell this.

February 27. Supped with Mr. Toynbee, who is one who lives in earnest, who knows in what moral rectitude consists and practises it. Heard some facts respecting W——; he seems to be or is a tyrant and a slave to money; he rejoices at the bankruptcy of a man who attempted to carry an electric telegraph from Dover to Calais. While men investigate science, rather for fame than for truth, they do not fulfil the duty that providence requires of them.

March 14. Had the pleasure of conversing for a brief space with Faraday this morning. He is not disposed to place faith in the magnetic experiments of Reichenbach, and says that, as of mesmerism, so he cannot believe in them until their law is found to be of invariable application, until they can mesmerise inorganic matter or a baby, who cannot be supposed to be a confederate. He has lost much time in the enquiry without any satisfactory results. He believes that gravitation will ultimately resolve itself into magnetism. What a true man he is, how pleasing it is to observe his recognition of the claims of his contemporaries in his present writings, even his subordinate receives honourable mention.

June 28. How much turpitude and want of moral principle I find in an individual with whom I daily work. May I regard his faults not to draw comparisons in my own favour, but as beacon lights to warn and guide me.

July 27. To work again, not at all irksome.

The past four weeks [a month's holiday in Shropshire and Wales] like a dream. After pleasure cometh pain. Yesterday heard of John's embarrassments, to-day my landlady asks for an increase of rent. Must work now and devise something new to increase my income that may enable me to clear off my liabilities and be in a position to assist others.

August 8. Received this week a most satisfactory and hope-inspiring letter from Walter at Sibford. The dear boy has reflected on what I have spoken and written to him, and understands that the great purpose of life is truth, and promises to aid his brothers and me in pursuing that object. I pray to be kept from all feelings in this and every other matter, except those that are devout and true.

August 17. Much conversation with Mrs. Merington yesterday on marriage and cognate topics. I think that, generally speaking, marriage swamps a man; he who has been active, persevering, energetic, suddenly sinks down into a digesting and sleeping machine. Oh, for the inner life which should make both man and wife lovers until the day of their death.

September 24. At the bottom of my purse, exciting cogitations as to how I should get over

the next four weeks, when comes a remittance
from Edinburgh (J. Hogg) ; and Charles Knight's
Mr. Ramsay called to give me some work
for the Almanack. Providence is very good,
may I reverently acknowledge it by striving to
make myself worthy of it.

October 6. Last Thursday I saw the fixing of
the Duke's Statue, but few persons present, no
huzzas. The world grows wise. How different
it would have been on the Continent.

October 28. Schönbein, the inventor of gun-
cotton, called with Mr. Barron of Stanmore.
They copied a paragraph from vol. 28, ' Silliman's
Journal,' recording the action of nitric acid on
vegetable fibre. Schönbein is about five feet
four inches in height, thick-set, round-shouldered,
short, sandy hair, shrewd, rustic, rather money-
getting face. He walked restlessly up and down
the office during the copying of the extract,
showing that his legs and name do not entirely
agree. Speaks English. On reading the paragraph said, "Ha, I think the man who knows
no more than that will never make gun-cotton."

Miss Mitford told me that when Bulwer was
sitting to Wyon for a medallion portrait, he objected that it was drawn rather stiff. "Not too
stiff to represent a young man," replied the

artist. "I thought you would have had the *Antinous* in your mind," returned the author of 'Paul Clifford.'

November 14. It appears from the delivery of Mr. Airy's address that he is somewhat to blame for the loss of the complete realisation of the discovery of the new planet by Adams of Cambridge, and there seems also to have been an attempt to make a Cambridge snuggery affair of it, for Challis and the Northumberland Equatorial. Airy confessed that some years ago he had declared, in answer to Mr. Hussey, that it was impossible there could be another planet beyond Uranus.

Mr. Ayres, the lecturer at the College of Chemistry, tells me some curious stories of Dr. D—— and his careless manner of lecturing; that when he gives a lecture on Botany the curator of the Botanic Garden is obliged to label each of his specimen plants, for the learned professor does not know them by name. If report speaks true, the facts and his paper on 'The Rotation of Crops' are decidedly at variance; one would say the Doctor's statement is not honest. Still it is hardly to be believed that he would publish what is not true.

November 25. To dine with G. Bubier. Spent

a very agreeable and intellectual evening. Our host told us that Perry's barrel or revolving inkstand was the invention of Bain, whose name is connected with the magnetic telegraph, and who was once a poor student in Aberdeen (so says Mr. Machonochie of Orsett). He being desperately pinched for cash, sat down one night determined to invent something that should be worth money before he went to bed. Having to tilt his inkstand to get sufficient ink, the thought of the improvement struck him ; he drew the patterns, went with them next morning to Perry's, who gave him £10 for the invention, with a promise of more should it succeed, and subsequently they sent him £100 in two remittances.

December 26. G. Bubier told me that once a meeting of some City men was held at the League Rooms in Fleet Street to nominate a gentleman to some vacant public position. Cobden was writing in the room in which they met. Opinions were loudly expressed, and all the persons named were rejected as not possessing the requisite qualifications ; we want a man with such and such qualifications, said one, enumerating a host of public virtues. "Gentlemen," said Cobden quietly, looking up, "the man that you want elects himself."

December 31. Last evening of 1846, a season for thought and retrospection. The past has been I think the most to me successful and improving year of my life. It has widened and strengthened the circle of my friends, increased and confirmed my knowledge, given me clearer perception of duty, and I trust a better capacity for performing it.

The daily blank that lay before me a year ago is now enlivened with a few flowers. I have the society of my first-born, which compensates in some degree for the absence of the others. My health is good and I have abundant employment, and reasonable hope of its continuance. During this year my sketch in the 'Penny Magazine,' the article in 'Blackwood,' and some of my most successful attempts at rhyme have been published. May I render thanks in sincerity and trust where thanks are due, and strive to make every advance in knowledge equalled by others in virtue and disinterestedness—to recognise life as a sphere of noble and ceaseless duties. To be patient and trust that where I am prepared for higher opportunities, they will not be slow to present themselves.

1847.

January 16. The special meeting of the Royal Society to discuss the " Legality " of the award of the gold medal to Mr. Beck is appointed for February 11. The result of that, as well as the new regulations adopted by the Council, will doubtless be to benefit the Society generally. I hear many opinions, some dictated by envy and jealousy, others out of spite, some by a desire for good ; alas for truth and sincerity.

The barrenness of my ideas, and the difficulty I find in generalising, oppress me sorely ; shall I never be able to write all I think and wish ? Would not a resolute course of study be beneficial to me—ought I not to persevere more in mathematics ?. My memory is much less perfect than it was. But I must not pause until my debts are paid. Heaven continuing to me health and strength I trust to be in a still better position by the end of the present year.

February 1. The new parcel post promises to

be a successful speculation, cheap carriage of goods and packages has long been a desideratum.

Later on in the same year, May 1, he writes—

The above opinion respecting the P. P. was fallacious. The Company has evaporated and given no sign to the creditors.

February 9. Much chicanery is apparent in various quarters respecting the "special meeting" on Thursday. Sir Jas. Clark said this afternoon that he is sorry he signed the requisition. Underneath the charge of *illegality* lurks the intention to force Dr. Roget to resign his office of Secretary.

February 11. The special meeting was held to-day, about 120 persons present. Mr. Stephens opened the proceedings by raising a question on a point of law, after Mr. Warren (author of 'Ten Thousand a Year'), then Mr. Broughton, and Mr. Grove. The Marquis spoke several times from the chair with much tact and temper. Altogether the discussion was characterised by great fairness, no attempt at quibble or conceal-ment. A forest of hands was held up in favour of the Council, *three* only for the opposition.

April 20. Yesterday completed my third year with the Royal Society; I pray to be enabled to persevere in my duties, that each succeeding year may find me better and wiser. It is rumoured that the Scientific Societies, ourselves among the rest, are to be removed from Somerset House to Burlington House.

May 8. Dr. Lee cannot forget his vexation about the award of the Royal Medal to Mr. Beck, and has written a letter on the subject to ' The Athenæum ' which the Editor declines to publish. He called to-day to enquire about his paper; is he again suspicious ?

I have been revolving Mrs. Merington's philosophy in my mind lately and find it very difficult to resolve. I cannot think that all manifestations of feeling should be suppressed for the sake of what is called genteel deportment —matter is to be sacrificed to manner—and a shake of the hand to resemble rather the meeting of tabooed fingers than any expression of cordiality. Under the name of not wounding the feelings of others, all feelings may be suppressed—every one is to do everything according to a precise set of rules, and society is to be stereotyped down to one dead level, without any show of individuality. Perfect decorum, it seems, consists in holding yourself aloof. Perhaps Cobbett was not far wrong, when he said that it was only persons with bad thoughts in their head who could object to a little sport. Perhaps if I live, I shall arrive at clearer notions on the subject in the course of a few years—at present I cannot think that either etiquette or money is the only thing worth living for.

August 2. At my post again, Somerset House, all in deshabille in consequence of repainting the rooms, the first time since thirty years. The paintings by Cipriani on the ceiling and coving of the ante-room will now be cleaned. On the 6th to the museum with Mr. Wesley from Burton— walking together in the evening and discussing some literary projects in which he requires my co-operation.

August 7. Finished copying Papin's treatise on the Vacuum, which has occupied me four days and is to be sent to Blois. Letter requesting two geological articles from Mr. R. Chambers—truly I ought to be glad and grateful for my prosperity. I pray for counsel and assistance to support and guide me on my way—that whether prosperity or adversity ensue my heart and conscience may be disciplined.

November 25. Discussion this evening at the Royal Society meeting; Barlow's supplement to paper on electro-magnetic currents read, and

Captain Johnson's on aberration of the compass
on board of iron steam vessels. Mr. Gravatt
began by saying that attention should be paid
to ascertain in what degree the compass was
affected while the steam was blowing off.
Colonel Sykes related a case of the action of a
compass having been inverted during the voyage
of the *Fame* iron steamer to India round the
Cape of Good Hope. On the way to the Cape
the South pole of the needle became the North,
but when sailing northwards again on leaving
the Cape it returned to its original position.
Another vessel went but the same anomaly
was not observed. Mr. Brooke called attention
to Dr. Scoresby's observation at the British
Association ; that the permanent magnetism of
a vessel might be altered by a series of
concussions from waves while on a given tack,
but that on another tack the same cause would
induce another magnetic state. Mr. Grove
doubted the truth of Dr. Scoresby's conclusion,
and thought that the induced magnetism would
depend on the position of the vessel with regard
to the magnetic meridian. Mr. Christie agreed
with Mr. Grove and suggested that the polarity
of the vessel might change after passing the
magnetic equator, and again on approaching it

on going up to Calcutta, but this left the fact
that in one vessel no disturbance of the needle
took place, unaccounted for.   All agreed that it
was a complicated and mysterious subject.   In
some instances the compass could only be
depended on when carried to the yard-arm.
Mr. Gravatt made some humorous remarks about
the uselessness of *small* compasses in a rapid
current of air, on a railway for instance, unless
the instrument be ventilated.

December 1. Hear to-day that Dr. Roget's
speech was considered egoistical, and in some
respects evasive.   Dr. Mantell said to me that
the anniversaries of the R. S. ought to exhibit
some first-rate eloquence, and masterly reviews
of science.   Grove would make an excellent
junior secretary.   This afternoon called on Barrett
in Mark Lane, who is to print the 'Family
Economist,' and gave him the material for the
first number.*

December 9. At the Council to-day a protest
from Wharton Jones and others was taken into
consideration—and not attended to—for the old
physiological committee was reappointed.   The

---

* The first monthly periodical of the kind.   It was
established by Wesley of Burton-on-Trent, and achieved
success.   Wesley became a successful bookseller in London.

result will probably be another special meeting. 'The Lancet' chuckles over Dr. Roget's secretarial obituary.

December 16. Dr. Whewell gave to-night the Bakerian Lecture, subject, 'Tides of the Pacific.' He gave a comprehensive and comprehensible sketch of his theory, and went on to show the impossibility of tracing co-tidal lines across great oceanic spaces. He inclines to believe in the theory of oscillation around a fixed centre. Let me say it in all humility, but to me it appeared to be a wilful shutting one's eyes to the truth. Mathematicians have so long persuaded themselves of the truth of the lunar theory that they will not willingly give it up. The Master of Trinity however honestly admitted the difficulties in the way, even to the tide which sets round Cape Horn, in the opposite direction to that prescribed by theory.

December 31. Last day of the year; dined and passed a pleasant evening with Mr. and Mrs. Toynbee. He gave me accounts of his visits to the various unions and workhouses throughout London. What scenes of dirt and degradation he has witnessed. He considers the masters generally more intelligent than the guardians.

1848.

January 22. Went to-day to inspect the
Central Electric Telegraph Office, an interesting
sight, my account of which is in 'Chambers'
Journal,' No. 217. The publication of my paper
on 'Walks to Office' affords me pleasure, as it is
one of my few attempts at original description.
What reason have I to be gratified for W. and
R. Chambers' continued kindness; it enables me
to pay my debts, and then I shall be a free man.
Heaven grant me strength and humility. G.
Bubier called yesterday, we agree in our approval
of Tennyson's new poem. He wishes me to
deliver one of a course of Lectures at Brixton.

February 9. At the Council meeting to-day
Sir H. De la Beche withdrew his motion for
restricting the time of the President's holding
office to two years. In consequence of Lord
Northampton's intended resignation it is proposed
that the Royal Society shall obtain additional
apartments and give soirées. There appears to
be some motive actuating the promoters of the

change, beyond that which manifests itself in their proceedings. Lieut.-Col. Sabine to-day in reply to Robt. Brown said he did not intend to be present any more at the Committee of Physics, as they occupy themselves with unimportant matters. According to Mr. Wheatstone the reason is that his papers on Meteorology and Magnetism in the Phil. Trans. for 1847, were not rewarded with the Royal Medal. What little things change men's dispositions. Professor De Morgan is angry because his paper on the Commercium Epistolicum is not to be printed. Sir David Brewster was estranged from the Society by a similar cause. Self and not science is the prime mover. When will philosophers be true to their calling?

March 23. Dr. Mantell's paper read this evening at the Royal Society supplies some omissions in Professor Owen's paper on the same subject, for which the Royal Medal was awarded some few years since. A warm discussion ensued ; first Mr. Christie grew angry because Dr. Mantell wished to read a short supplement. Mr. Owen spoke at some length, throwing discredit and contempt on the whole paper. Mr. Bowerbank in favour. Then Dr. Buckland in a most luminous and humorous discourse, then Dr.

Carpenter inclining to Mr. Owen's view. Mr. Gray made a few remarks, and at past eleven o'clock Dr. Mantell replied. The most interesting meeting which I have yet known at the Royal Society. Mr. Owen seems to be unnecessarily severe. He however does not like to have his own views contested. R. Waldo Emerson was present at this meeting; unfortunately I did not know it until the next day.

May 24. Heard Captain Smyth relate the following anecdote. Mr. Babbage had a title engraved for his ninth Bridgewater Treatise. The drawing was by Corbould, and it represented a passage in the Arabian Nights where some one drawing aside a curtain says, " You have seen eight marvels, I will now show you a ninth more beautiful than all the rest." Mr. Babbage had quite overlooked the arrogant assumption ; it was suppressed by the advice of a friend.

June 19. Out riding with Lovejoy [a well-known bookseller at Reading] and Miss Mitford. The latter entertains but a poor opinion of Lamartine and condemns Tennyson's 'Princess.' Saw a letter from Mr. Browning. Talfourd is about to publish 'Final Memorials of Charles Lamb,' in which many extraordinary particulars will come out. G. Lovejoy hears from Charles

Tilt that Dickens' 'Pickwick' was not at first popular. The work had been offered to various publishers and Chapman and Hall were not over pleased with their bargain. Tilt sold 1200 of No. 6, and the publishers sent to Dickens a cheque for £30 over and above the £8 per sheet agreed on; he acknowledged it. For No. 7 they sent him an extra cheque for £60 which he did not acknowledge. For No. 8, a cheque for £100 which he returned. They altered the one into four and then the author kept it. Altogether he received for Pickwick £1200 more than was stipulated for.

July 25. Called on Mr. Sylvester, F.R.S., for his subscription due to the Society—he took offence at what he called my "peremptory prescribing of a time" in which he should pay. An atrocious mistake, to call it by no worse name, on his part. He insisted also that I should address him as "Sir" with every sentence, or query, or rejoinder. What a paltry vanity.

September 17. Quetelet says in his 'Faits Soriaux' that it would be well if all families would keep a book in which to record everything that causes them joy or sorrow. Also the growth and development of their children; were this followed up we should have an ample register

of human joys and sorrows and perhaps be able
to limit the scope of human error.   The weather
has been charming the whole of this month.
News of the insurrection in Frankfort on account
of the Schleswig armistice.   Lord G. Bentinck
died suddenly.   Le Verrier's planet is said not
to be a complete discovery.   Discussion concerning it in the French Academy.   Claim set
up by M. Moigno to have one of Wheatstone's
newly-invented Polar clocks fixed to the Garden
of the Luxemburg, where Malus made his
discoveries of polarization.

November 2.   Our first Council for the season ;
the reformers have carried their point, and got
their own men nominated for election on the
30th.   Whatever changes may occur I scarcely
think that I shall lose my post.

Called on Mr. Chadwick in the morning ; with
what different feelings I went to his house for
the first interview nearly five years ago.   He is
determined to do his duty in the New Sanitary
Commission, and to appoint no officers but such
as are qualified and will do their duty also.   He
tells me I am quite right to bring my boys up to
trade.

November 7.   I am directing our annual council
lists.   Mr. Airy and Horner are new members,

doubtless that the latter may assist the progressive movement in the Society. Mr. Grove has taken Siddons' House in Upper Baker Street. Young Brayley tells me that of all the patent cases which he (Grove) undertook to defend at the bar he never lost one. Last year in directing the lists I omitted " Bart." after Sir John C——'s name ; he came shortly afterwards to the Library in a rage, asking if it were to a dirty Knight. This will match with Mr. Sylvester's asking for the designation "Sir."

November 18. Much indignation has been excited by the New Council list. The physiologists contend that one Secretary at least should be of their body, instead of as proposed two physical science men, Christie and Grove. Active canvassing on both sides. Mr. Thomas Bell is the physiologists' candidate. Bowerbank and Bishop are two of the most active in his behalf. On the other side Mr. Gassiot and Sir Charles Lyell. Grove, though proud as Welshmen are, is perhaps best qualified for the post. Some of the physiologists will support him because they disapprove of what has been done by the physiological committee. At all events a debate in our meeting room will shake some of the philosophical dust from off the walls.

November 25. Mr. Weld went to Hull to-day. 'The Athenæum' has seven letters on the forthcoming election at our Royal Society anniversary. The contest threatens to become more and more keen.   Mr. Sheepshanks, Robert Brown, and Dr. Roget came in in the afternoon, and conferred on the proceedings to be taken.   Mr. Sheepshanks means to snub the lawyers, and has primed himself from the original charters. Will science or faction conquer?   There have been many secret conferences this week—much trimming and time-serving — alas for human nature.

November 30. The eventful day, the ballot began.   Mr. Faraday made some remarks about the list.   Sir H. De la Beche cautioned the company to look well at their lists before voting, wishing, as he said, "all to be fair and above board."   Mr. Horner spoke also deprecating the introduction of Dr. Bell's name in the House list.   Then Sir Charles Lyell in a speech explained why Mr. Grove had been put forward, and advocated his claim.   Then Mr. Wharton Jones, after whom a ballot.   The Fellows went up on one side in single file, delivered their papers and out at the other.   The scrutators, Dr. J. R. Bennett and Dr. Royle, proceeded to their scrutiny in the Council Room, and in their absence Sir R. H. Inglis moved a vote of thanks to the President in a laudatory speech, seconded by Mr. Broughton in the same strain. Dr. Paris then moved thanks to Dr. Roget for his twenty years' services, seconded by Baden Powell. Mr. A. J. Stephens then spoke, iterating his "Corporation law" once more. The meeting proving tedious, Lord Northampton gave up the chair to Mr. Rennie, and with a number of the Fellows went away to dinner at the Freemasons' tavern. Presently the scrutineers came in and reported that with Mr. Bell as Secretary, the House list was elected. The numbers were, for Mr. Christie 215, for Mr. Bell 134, for Mr. Grove 108. After this declaration Mr. Rennie inducted the new President, Lord Rosse, to the chair. His lordship adjourned the meeting to this day se'nnight, and the great event was over. It would have been more graceful had Lord Northampton stayed to induct his successor.

December 1. This morning I hear that one of the cross lists that came before the scrutators yesterday bore for President, "Sir Roderick Impey Murchison, Knight, Emperor of *all* the Russias," a joke played off by some one who knows Sir R.'s weakness.

1849.

January 1. Mr. Wheatstone said to-day at the Library, that Dr. Holland, who had just seen Arago at Paris, that the great astronomer affirms there will be yet another insurrection, and if so, he hopes he may fall in it. To Meurice, a Swiss gentleman, he said he would depart for America if things do not go on better. Thus a great philosopher is either a-weary of life or misanthrope.

January 18. Mr. Glyde says that John Walter has sold 'The Times' to one of the firm of Jones Lloyd—hardly credible.

January 22. Letter to-day from Gainsborough. R. Thompson proposes terms (for his son Walter's apprenticeship). I have replied to . . . . Fred went to Croydon on Sunday to visit the Friends' School, it is very probable he will be admitted as a teacher. I shall thus have reason to be thankful if two of my boys obtain situations with friends. Soon my darling Harry will go, and then I shall be alone again. Mr. Manby tells

Weld that Jenny Lind has married Lumley, it is to be publicly acknowledged by the end of the year.

February 7. Yesterday saw T. Babington Macaulay at our evening meeting; he has an intellectual forehead, grey hair, wears a white neckcloth which gives him the appearance of an Independent parson. He is square built, about five feet eight high, but with nothing particularly impressive about him. Rather quick and abrupt in speech, clear voice, words seem to come from him as pellets from a tube. At Sir R. H. Inglis' request I showed to him, to Lords Mahon, Dalmeny, and M. Van de Weyer, the Belgian Minister, our relics of Newton. Lord Rosse was in the chair for the second time. Samuel Warren was at the meeting, he says Lord Rosse is "a muff," and that he (S. W.) is to be made Q.C.

February 10. Mr. Moseley said to-day that if all the *On dits* of science, etc., were set down daily, the result would be valuable some twenty years hence. He mentioned that at the desire of the Bishop of Sodor and Man, all the clergy of his diocese keep a diary of all that happens in their respective parishes.

February 20. Weld told me of a *bon mot* of Croker's. When the Athenæum Club was first founded Croker was urgent that no man should be admitted who was not distinguished in literature general or learned. Soon after he proposed the Duke of Wellington, when some one said, "The Duke has never written a book." "True," replied Croker, "but he is a capital hand at reviews."

March 16. Robert Brown gave us a story to-day about J. E. Gray, of the British Museum. A few days ago Gray was in an omnibus opposite a lady (Irish) who had a fat lap-dog in her lap. "Madam," said Gray, "you feed your dog too much." "Indeed, sir," was the reply, "then I only do for my dog what you do for yourself."

May 8. Went to the sale at Gore House on Saturday, a luxurious and voluptuous display. Yesterday the silver model of the Countess' [of Blessington's] hands sold for £40.

November 5. Dr. Mantell intends to claim the Royal Medal for his 'Palæontological Papers.' I think he ought to have it, but I do not like to see a philosopher so anxious for a mere medal.

November 8. Am reading Paget's 'Hungary,' 'Newman on the Soul,' and Emerson's 'Second Series of Essays,' the two latter most suggestive books. Have commenced writing a 'Catechism on Sanitation' for Wesley, my first attempt in that line. My lists now show that I have written two hundred articles for ' Chambers' Journal' since 1844; what a change in my ability and circumstances in that period. Then imperfect powers and almost penury—now tolerable command of language in a literary composition, with an income sufficient.for all my wants, and good prospects, the most cheering of which is the almost certainty of being able to provide well for my little boys. May my plodding habit continue, and may I feel properly grateful for all mercies.

November 22. Our first meeting for the session, a good attendance and of eminent men. Sir John Ross looking well and hearty after his Arctic adventures. Sir R. Murchison, Sir R. Inglis, Sir F. Pollock, and others. T. B. Macaulay was elected a Fellow.

February 29. Heard to-day that *Erebus* and *Terror* are to be re-equipped, and that Sir Ed. Belcher is to command the next expedition to go out in search of Franklin.

November 30. Our anniversary held this day. Lord Rosse delivered his address; although superior to what have been spoken by Lord Northampton, it disappointed me. The sketch of scientific progress was too meagre. Weber's ' Electro

Dynanometrics' and Deve's 'Isothermal Lines' were the chief topics. Owen's paper on 'The Development of Chelonian Reptiles' is said to be a plagiarism from Rathke's work on the same subject. Sir R. Inglis as usual moved thanks for the address and that it be printed : but I thought that his observation that "the address had never been excelled in that meeting" rather fulsome. Sir R. afterwards moved a vote of thanks to Lord John Russell for his proposal to place £1000 annually at the disposal of the Council, for the purpose of assisting scientific men in their investigations, as had been stated to the meeting by the President in his address. Sir R. took occasion to say that he looked back on a proud hour of his life, that on which he occupied the chair, and had the honour of admitting Lord John into the Society. The worthy baronet would certainly be more effective in his speaking were his style less involved and wordy. The Copley Medal was given to Sir R. Murchison for his researches in Geology ; one Royal Medal to Col. Sabine for his paper on 'Terrestrial Magnetism'; one to Dr. Mantall for his paper on the 'Iguanodon.' No opposition was offered to the House list, and the meeting passed off quietly.

December 6. At our meeting to-night T. B.

Macaulay was admitted by Mr. Rennie, who occupied the chair as Vice-President.

December 28. Dined with Mr. and Mrs. Toynbee—conversation on rewards of labour—the consciousness of doing one's duty is a noble recompense. Those who struggle most generally the most successful. Sir B. Brodie says it was best for him to have to fight his way. Dr. Hunter and Dr. Baillie to wit; the latter took life easy, thinking the former would leave him his property. But Hunter one day said—"I had so much pleasure in earning my money, that I mean you to have the pleasure too. I shall not leave you a penny." Dr. Baillie became a most successful practitioner.

Our Working Classes' Committee meeting at eight o'clock, a good attendance. Heard of J. Cassell—came originally to London with 3*d.* in his pocket. Knows how to make money, but is said to be honest and liberal in his views. He spent £1000 in advertising ' Standard of Freedom ' before a number was published.

All this year and for a good many succeeding years White was harassed with domestic troubles. His wife had left him and was living separate, his sons as they grew up and embarked in different trades had the same love of change of occupation which he had suffered from in his youth. Though he was the most affectionate of fathers, his life from 1849 is a long record of domestic or family trouble, and of excuses for misdoings. His "brave grocer boy" ran away from his employer, but the father had harder words for the master than for the boy, and still has the excuse ready (*November* 8, 1849) : " I must confess that I cannot condemn a bold self-asserting feeling in a boy, when not accompanied by sneaking dishonesty. Some boys are good only because they have not spirit or energy to travel out of the prescribed track." It is small wonder that his sons very quickly took matters into their own hands, and a few years later, to his grief, emigrated to America and then to Australia, in search of pastures new, leaving him to " find my mind and heart drawing contrasts between loneliness and companionship, and fancying the pleasures of knowing

'That an eye
Will watch for my return and beam
The brighter for my coming.'"

1850.

March 15. In the evening heard the Astronomer Royal lecture at the Royal Institution on Terrestrial Magnetism. A most interesting discourse; lasted nearly two hours. Prince Albert in the chair. Mr. Airy does not believe there are two poles of magnetism in each hemisphere—he explained the instruments, described Barlow's Globe, Christie's plate with its opposite effects of magnetism, Hausbeen's supposition of two magnets 2600 miles long. Gauss of a force of magnetism in each cubic yard of earth equal to eight magnets of 1lb. each. He described tracings of magnetic disturbances, and stated that to take one of these, trace it to its source, discuss it in every possible way, should most likely lead to a determination of causes. His recapitulation was not well listened to owing to the impatience of the audience. His commencement was excellent—defining science as not a knowledge of facts (even organised facts), but as knowledge of causes, the mechanical causes.

H

Illustration: astronomy which had a theory for every celestial movement, till Newton came and referred them all to the mechanical agency of gravitation.

March 28. My paper 'Sanitary Movement' published to-day, an event which gives me pleasure. Publication forms one of the gratifying periods in a literary life.

April 3. To-day had the satisfaction of posting 'Arctic Explorations' to Messrs. Chambers. This was the second "Paper" I had written [for 'Chambèrs' Papers for the People']. The subject proved more difficult than I expected, arising chiefly from the voluminous materials to be condensed.

April 15. Letter from Mr. Page speaking favourably of my writings. The 'Arctic' is to commence 3rd vol. of the "Papers"; they would like to have 'Antarctic' to lead off vol. 4; but under all circumstances I must defer this until after July, as I have much to do in preparation for my annual holiday.

May 1. Long chat with Mr. Sheepshanks. He has now discovered the means of getting a true thermometer. If the freezing point be made after the tube is boiled it will not long remain true. It will be $\frac{3}{10}$ of a degree lower.

This point must be marked before boiling. He spoke of Airy's discovery of the compensation for the compass of iron ships. The first iron vessel brought round from Liverpool to the Thames was nearly lost; on one occasion she was running straight for the south coast when her captain was warned by a fisherman. He (the captain) although a good seaman became so nervous and anxious that after he arrived he declared he never would go to sea in an iron ship. Airy compensates the perturbation by placing a magnet near the compass, which balances the permanent magnetism in the vessel and leaves the free magnetism to act its usual effects. The Steam Navigation Company invited Airy to a pleasure trip in a vessel thus compensated, and he was gratified to hear the captain above alluded to, as he noted the bearing in each reach of the river, express entire confidence in the contrivance.

May 27. Lord Brougham has been several times at the library lately respecting his paper, and his having stated that Arago said the Copley Medal was disgraced by having been awarded to Sir Roderick Murchison. The latter wrote to Paris to ask the Frenchman if the report were true. He denies emphatically and observes of Lord B.—
" Quand cet homme cera-t-il d'être un étourdi."

1851.

October 30. At noon to-day Kossuth passed along the Strand on his way to the city—he was in a barouche drawn by four grey horses, followed by two or three similar vehicles. A crowd which lined both sides of the street as far as eye could reach began to collect before eleven o'clock. At two of the newspaper offices, the Hungarian tricolour was displayed in a large banner of green, red, and white—and " Welcome to Kossuth." What hearty shouts and greetings were sent up when the hero came in sight : the carriage was frequently stopped, and numbers of working men thrust out their hands, which the Hungarian shook with great good-humour. The enthusiasm was genuine, a cordial recognition of fair play, of a man of the people. All went off pleasantly, notwithstanding the throng and the temporary stoppage of street traffic. Kossuth sat behind at one side, I got a good view of him: he has brown hair long and flowing apparently, and wore a black low crown hat. He returned at 3 p.m. in the same way.

December 8. Heard from Mr. J. S. Graves to-day that Sam Warren got £1000 from Blackwood for the ' Lily and the Bee.'

December 19. After our meeting last night Col. Sabine told me several anecdotes of his sojourn in the Arctic regions when he was out with Captain Parry on his first expedition. They used to eat all the animals they could shoot or catch; amongst others the wollymokes or petrels, which he says were the nastiest things imaginable; still they ate wollymoke pies until one day some one was made ill by them, and that put an end to them.

When they were out walking on Melville Island, if ever they found a small lump of coal they carried it off in their pocket to warm the water for their evening glass of grog.

The greatest unanimity prevailed among all hands, from high to low: the only approach to a disturbance being a tipsy middy who had been invited to dinner from the other ship. Very different this from the quarrelsomeness of the Austin expedition.

Parry was born in 1788, and was at that time about thirty years old; he and the Colonel were in the crow's nest as they passed Wellington Channel, and they had nearly decided taking that passage when Parry remembered he was to go to the North-West.

What an interesting book might be written of the domestic history of this expedition, of all that was said and done as it were in private life. What would we not give for such an inner history of Baffin's or Hudson's expeditions!

1852.

January 26. Spoke to our treasurer to-day about the foreign parcels delivery to be undertaken by the Royal Society, and he rebuked me somewhat sharply for entertaining notions on the subject differing from his own. It is one of those rebuffs, galling for a time, which subordinates have to endure. However, I have done my duty and the rest remains for time to decide.

January 29. Circular came to the office to-day, announcing a meeting of gentlemen at the Thatched House to deliberate on offering their services to Government in case of danger from the apprehended invasion by the French.

February 12. To-day Sabine, Bell, Owen and Playfair met as a committee to discuss the question of removal of Royal Society from Somerset House. A Palace of Science was mentioned, but there would be an objection to amalgamation of societies though not to juxtaposition under one roof. Lord Rosse is to see Lord John, and if no better can be, an attempt will be made to add the rooms of the London University to ours, and we shall remain where we are. The latter would please me best.

February 27. It is believed that Lord Rosse's intimacy with Earl Derby will facilitate the getting of a mansion for the Royal Society.

March 2. Dined at Toynbee's. Was introduced to Dr. Guggenbühl and Liebermann, the former a cheerful-looking man, candid open eyes. He has been a teetotaller from his birth. He has been making a little tour in Italy to propagate his views as to the cure of goitre and cretinism, and says there are many more cases of both in England than is commonly supposed.

March 11. Our president came to town. He will not see Lord Derby as yet about a mansion for the Royal Society, not only because the Premier is busy, but also because he cares nothing for science and makes horse-racing his recreation. The project, however, for getting a mansion, perhaps the present National Gallery, is still vigorously discussed in the Philosophical Club, as also the making the Royal Society the grand recipient and distributor of donations of scientific books from all parts of the world.

March 22. To the St. James' Theatre at two o'clock, heard Fanny Kemble read ' Midsummer

Night's Dream' with Mendelssohn's music. The first time I ever saw the lady, she appears to be forty, and to have a touch of the devil in her, she however read the play beautifully with perfect command and intonation of voice, and perception of character.

May 6. Our Government Grant Committee met to-day. While at G. Merington's this evening a proposition was made to me to go out to Australia, to help them form a little colony with the Merington family.

June 19. Left London at 2 p.m. with G. Merington for Fulbourn, where at seven o'clock I gave a lecture on emigration to Australia to a congregation of labourers.

August 5. Commenced to work straightway upon ' Rising and Thriving,' for W. Wesley, and completed it in about ten days.

Now a complication of malady and malaise has seized me, with greater and more continuous pain than ever I felt. I cannot work, scarce can I eat or sleep.

September 17. My health happily mending. . . . Was more struck to-day than usual by the remark that there is more significance in the dry details of daily life than we think. Hawthorn repeats it in the introduction to the ' Scarlet

Letter.' Why cannot we look on and note them as they pass with the same perception of their meaning as we have when we look back on them ?

November 13. Went to see the lying in state of the Duke [of Wellington], breakfasted early, walked to Chelsea, arrived as one hundred of the Guards were marching in to the relief. Thrust myself into the crowd, where the pressure was tremendous. Held my ground, though the shrieks of the women were at times terrifying. In an hour and ten minutes I had passed through and was out again. The spectacle one well worth seeing. The crush increased as the day grew later, and two women were killed and several persons injured.

November 18. Rose at a quarter to six, left home before seven, having to get early to Somerset House ; difficulty in crossing Holborn [the day of the Duke of Wellington's funeral], which was thronged with carriages, the traverse only effected by aid of the police. The same in the Strand, but the press was nothing like so great as I had expected. Found our platform outside the window complete, ample room and as snug as a private box. There were our Secretary and Treasurer and the Poet Laureate ; had much talk with the latter.

1853.

April 2. Weld tells me that Prof. O. says that Robert Brown is one of the greatest bars to scientific progress that we have now to contend with—over-cautious and jealous. W. states also that he is thinking of buying a government annuity and retiring into Devonshire in a few years' time.

April 23. My 42nd birthday. My literary capacities improve, but my spiritual and moral nature remain too lifeless and weak, and need a keener sense of my duties to God, and such a fear of Him as shall incline me to the good and keep me from the evil. It is the loving fear, not the base fear that I want. A heart ready to sympathise with sorrow and rejoice in the virtue of others. My dear boys at Reading sent me cheerful holiday letters, and Harry a few loving verses which brought the tears into my eyes, and made me feel with shame and remorse how much I am below the standard which the dear lad assigns to me. However, grant me grace to. ask in sincerity for His help.

May 2. Mr. Wheatstone told me the Marquis of Lansdowne went to a pre-private view of the Exhibition last Friday and marked thirty pictures for purchase, but found on enquiry they were all sold. Is not this a sign that artists are encouraged? The Manchester merchants are said to be great purchasers of good pictures. Mr. W. has engaged himself as scientific referee to the United Kingdom Telegraph Company. Salary £700 a year for three years. They will lay a pipe underground from London to Edinburgh along the turnpike road, which is to contain 100 wires with branches. Part of the scheme is to send shilling messages, and to let one or more wires exclusively to any parties.

He has been to-day to see the Prince's Library under the care of Dr. Becker at Buckingham Palace. There is a large collection of comic works, including 'Punch' and his continental congeners.

May 19. Mr. Bell came to the office this afternoon and took occasion to express so favourable an opinion for himself and Col. Sabine of my services, as quite gratified me. He promised also to take measures in due time, at the close of this year, to get me an advance of salary. This will be a real encouragement, but it makes me reflect with some shame on my many short-comings.

Mons. Ville from Paris told me two anecdotes of Laplace. Once Napoleon said to the great mathematician—"Mons. Laplace, j'ai lu votre 'Mécanique Celeste' avec grand plaisir; c'est très bien raisonné et c'est un ouvrage qui fait honneur à son auteur et au siècle; mais je remarque que vous servez pas une seule fois du nom di Dieu." "Sire, je n'avais pas besoin de faire cet hypothèse," was the reply While on his death-bed and in the last half-hour of his life, Laplace taking up a volume of his 'Mécanique Celeste' said—"Tout ça ce ne sont que des blagues. Il n'y a de vrai que l'amour."

September 8. The obituary of 'The Times' records the death of my friend Jas. Simpson of Edinburgh . . . he was truly a friend to me, for it was by his introduction of me to Mr. Chadwick that I obtained the situation I hold at the Royal Society.

1854.

February 10. The Council sat to-day and
sat late. Just before breaking up I was called
into the room, when Col. Sabine, the chairman,
stood up and addressed me thus—" Mr.
White, I am happy to inform you, that in consideration of
the zeal you have shown in the service of the
Society, your courteous manner to the Fellows
when they come on business, and the promptitude
with which you discharge your duties, the
Council have raised your salary to £150 a year."
This was very gratifying; I could not reply as I
wished, but hope to prove my thanks by deeds.

May 21. Here is another book [volume of
diary] begun. My last was commenced in New
York in July 1838, when I worked with the
plane and not the pen, and when I little expected
to be other than a handicraftsman all
my life. I am changed since then; in position,
in views of life and its duties: have seen some
trying vicissitudes; but am thankful to be able
to say have ever kept up a good heart of hope.

This present book will last me half a score of
years, when, if I look back in life, I may see
other changes—much to regret, some things to be
glad of. If we could only be as great as our aspirations! The time finds me well in health and circumstances, still willing to work, and with a relish
for it. Four of my boys are at Reading—Fred
teacher at his uncle's—Hamilton and Tyrone at
school, and Harry apprenticed to Huntley and
Boome—a few months more and it will be
Hamilton's turn to begin the battle of life, and so
they emerge one after the other from boyhood,
reminding me that years are accumulating on my
own head. Who knows how soon their number
shall be told! Does this reflection affect me as it
ought? Oh! this problem of life, why is duty so
hard in aspect? Why do the young seem to me
less trustful and earnest than when I was a boy?
less impressible by advice. Is it that as we
advance in years the past ever looks best, or that
the generations grow up with less veneration
than formerly? If we could but carry out the
Gospel precepts in their integrity the problem
would doubtless be solved.

October 21. Mr. Wheatstone to-day introduced Mr. Ruhmkoff to me, who had come over
to sell a newly invented electric telegraph which

prints sixty words a minute with great facility. He (Mr. W.) showed me some sheets with columns of figures, which are composed, printed, and stereotyped by a calculating machine invented by a Swede gentleman, G. Scheutz.

November 16. Our first meeting. Scanty, owing probably to the cards having been sent out only yesterday. The delay was occasioned by Lord Rosse, who retires from the Presidency, being reluctant to have his name in the Council list. He has however yielded, though not without an argument in favour of his own views. Mr. George Scheutz, the inventor of the calculating machine, was present.

November 30. At our Anniversary Meeting to-day Lord Rosse resigned the Presidency after six years of office. His farewell address contained some home truths about the defective system of Education at Oxford ; but was spoiled by his lugging in a eulogium of Mr. Babbage's calculating machine. His lordship does not retire quite so comfortably as could be wished. In a letter to the Treasurer, he says he " has been a puppet in the hands of the Council," and he has written some hard words to Mr. Weld. At present his successor, Lord Wrottesley, is not very popular.

1855.

January 19. Mr. Wheatstone says he has invented a new system of electric telegraph so simple and cheap that it must supersede all others, but that his engagement at £700 a year with the original company to give them all his discoveries, prevents his making it public. He does not wish them to benefit by it.

Dr. H—— says that Robert Brown has kept botanical science in England back a century, by his extreme caution and unwillingness that new men with new knowledge should come into notice. Corroborative facts will appear when the old man dies, and his dusty stores in Soho Square are rummaged. He saves nothing of his £800 a year, and is said to give much away in private charity. He lives in one room among piles of books, and every evening at six may be seen entering the Athenæum to dine.

February 2. James Heywood reports that the plans of the rooms and halls destined for the Royal Society in Burlington House are ready.

Our Treasurer and Secretaries held a conference thereupon, and Mr. Weld went down to the Board of Trade to see plans, and was told there were none in existence. So much for rumour.

February 20. In the evening to Parker's, the bookseller, to his weekly gathering of literary friends and gossips. Went at about a quarter to ten, only four present when I arrived. Mr. Clough, an Oxford tutor. Afterwards came Robert Bell, editor of the edition of 'British Poets,' a rather coarse-looking man, who does not believe in the immortality of the soul, whereon I joined issue with him. R. C. French was expected, but did not come.

February 23. My 'Walk to the Land's End' has now been refused by three publishers. Well, must try again. I think it is in me to write a book, and with the blessing of Providence will persevere.

March 13. Left my book for consideration at Smith and Elder's.

March 17. My MS. book back from Smith and Elder's. They like it, but don't think enough purchasers would be found to make it worth their while to publish it. Must try elsewhere and not lose courage in the meanwhile.

March 19. Left my MS. with Chapman and Hall.

May 3. Stayed at home this morning, finished my manuscript, and went forthwith and left it with Chapman and Hall. I am to have an answer next Wednesday. Chapman repeated that the first portion was " liked very well." The writing has been a hard though pleasurable task. I have striven to make it the best book on the subject yet; to please myself as well as others. I ought to feel thankful to God for the ability He has given me, and to show my thankfulness by a diligent and virtuous life.

June 27. Chapman gave me a cheque for £50 in payment of the copyright of my book. . . I feel glad and grateful at my success.

August 19. To Snowden's. Saw a daguerreotype of Thomas Carlyle, said to be a good likeness, but very grim. Also two of his wife. The latter, with Miss Jewsbury, are to see Mr. Snowden, so that Miss J. may see a specimen of a kind husband, and then decide to marry—if she will.

1856.

February 14. Mr. Babbage called to make
enquiry about the notice of him I have written
for Knight's ' Encyclopædia of Biography.' He
stayed an hour talking of many subjects, prayer,
philosophy, religion. He says he believes in God
and woman, not in priest or devil. That a man
who thinks and knows anything is intimately
acquainted with the greatest fool on earth. . . .
He says he is happy because always intellectually
employed : that he has great works in hand, one
of which is a Philosophical Dictionary superior
to any that has yet appeared. If he could have
his life over again, he would not change any
of its incidents.

March 8. Thos. (Carlyle) thinks we are wrong
in having begun war with Russia, that Russia is
a growing power with a master that knows how
to govern. That Macaulay is a mere commonplace writer, one who just puts ordinary thoughts
into fluent language, hence his popularity.

October 26. There was a party at T. Carlyle's

last Monday. Browning present, who told a story of R. H. Horne all bearded having met Fox in Piccadilly. Carlyle says Gilfillan is a brute, a wild ass's colt.

1857.

April 30. Our last meeting in the old Meeting-room at Somerset House.

May 4. Nothing now of ours left at Somerset House except pictures in the Meeting-room.

May 7. Our first evening meeting in the New Hall [Burlington House].

1858.

May 8. Kingsley came to Burlington House to look at the drawings made by Atkinson. . . I liked Kingsley's appearance and manner.

May 20. Our Council sat to-day to consider who shall be the new President, as Lord Wrottesley intends to resign next November. His lordship, Mr. Grove, and Mr. Gassiot went as a deputation to ask Mr. Faraday to allow himself to be put in nomination. He requested a day to consider, but the impression is that he will refuse.

The Lord Chief Baron came to hear his paper read in the evening. While looking out at our court-yard, I told him what Walpole had said about the colonnade. He said he was fourteen when Walpole died in 1797. I told him he did not look so old. He took off his hat to show his grey hairs, and told me he rises at three in the morning, goes to bed at nine, and attributes his good health and activity to that. He began early rising when considering the Bridgewater case, viz. the fortune left on condition that the legatee should obtain a peerage—on which the House of Lords agreed with the minority of the Judges that the condition was an illegal one. His lordship found that he could get through the work so comfortably and with such economy of time, that he continued early rising. "I light my own fire," he said, "and make a jolly cup of tea."

August 29. With T. Wheeler, Mr. Fletcher, and two others to Beckenham, where we heard Miss Marsh preach a sincere but weak sermon in a barn. Did not think the trip worth all the time it took us.

September 21. To Grantham in charge of the Royal Society's Newton Telescope. A pleasant journey, and all new to me. The telescope was carried in the procession by some of the Grammar School boys. Lord Brougham delivered the oration to inaugurate the Statue of Newton. It was a rare intellectual treat. A *déjeuner* afterwards in the Corn Exchange, about 400 sat down. Lord B. made a short speech, followed by Professor Owen, Bishop of Lincoln, Sir B. Brodie, Sir E. Cust, Monckton Milnes, and others. Back to London by 10 p.m.

October 18. Lowe, editor of the 'Critic,' called to ask me to write a weekly article, or Summary of Science. Declined. Don't like task work.

November 30. Sir B. Brodie elected President at our anniversary meeting to-day. Lord Wrottesley awkwardly enough did not induct him, but left him to instal himself in the Chair. At the dinner afterwards (so Sir C. Lyell tells me), Jamin, who had received the Romford Medal, made a very neat speech, and mentioned about the three who had had it, Biot, Pasteur, Fresvel. The latter, then on his death-bed, said it would be a crown for his tomb. Then Jamin said, "After these three illustrious men what must my emotions be on receiving the medal—I whose achievements can hardly have been said to have begun?" Sir C. Lyell is greatly pleased at having the Copley Medal; and said that when twenty-five years ago he had the Royal Medal for his 'Principles,' the elder Murray, having an eye to business and a second edition, wished to get a literary dinner in honour of the event. "I being green at the time," said Sir C., "and not liking the idea of meeting a number of friends, to be praised, threw cold water on the dinner, and it was given up. But I know better now, publicity is what a man in my position wants. Murray, who, when sober, sometimes repented of the bargains he made when metaphorically drunk, told me my 'Principles' was the first scientific he had which made money. It produced income to him and to me, not a large sum, but enough to enable me to travel, and that was what I then wanted."

December 22. Mr. Faraday called to enquire on the part of Sir Walter Trevelyan whether a MS. of Meteorological Observations made in Greenland would be acceptable. The question answered, I expressed my pleasure at seeing him look so well, and asked him if he were writing a paper for the Royal. He shook his head, "No, I am too old." "Too old, why age brings wisdom." "Yes, but one may overshoot the wisdom." "You cannot mean that you have outlived your wisdom." "Something like it, for my memory is gone : if I make an experiment I forget before twelve hours are over whether the result was positive or negative, and how can I write a paper while that is the case? No, I must content myself with giving my lectures to children."

1859.

June 23. While in Chapman's counting-house
was introduced to Thackeray, who happened to
come in. Had heard so often that he was ugly,
that I was agreeably surprised to find him otherwise : he has a lively eye, fresh colour, and
an appearance of old youth or youthful age.
Told him I had been the means of making many
persons like his books, and longed to tell him
that he had harped too much on the sentimental
string in the 'Virginians,' to the exclusion of
incident and the detriment of the work. He said
he wished he had five numbers yet instead of
three. In reply to a remark of F. Chapman's he
said that if he had a rich uncle he should strangle
him. Then F. C., "You say that who can write
such books ; why, if I could write such books
as yours I wouldn't envy even Rothschild. I
don't as it is." Soon after he rose, shook hands,
expressed pleasure at having made my acquaintance, and said, "I go away a little taller, Mr.
White, for this conversation with you." During

the conversation F. C. said that E. Chapman had once said to Dickens, "Take a pinch of snuff," and handed him a box containing £1400.

September 14. Read a letter from Mrs. Carlyle, dated Aberdeen, says she has all that is essential to happiness, except the faculty of being happy.

November 1. Received a circular from Smith and Elder enclosing a printed one from Thackeray, inviting me to join the staff of writers for the new periodical ' The Cornhill Magazine,' promising a handsome honorarium.

September 22. Captain McClintock called to deliver in his report of the scientific researches made for the long-lost Franklin Expedition. Sir John died in 1847, the ships were abandoned in 1848 : the track of the crews was followed to Montreal Island, where the last survivors must have perished miserably. So Lady Franklin's final search has solved the terrible mystery.

November 17. Our first meeting for the Session. McClintock's report of his scientific researches during his Arctic search, followed by a discussion of the magnetic results by General Sabine, in which the means of fixing the magnetic pole were treated of, the phenomena of auroræ, discriminating those due to luminous mists, and the connection between the sun and magnetism,

as shown by Carrington's observation from his
observatory at Redhill, by a sun-spot of intense
brightness for ten minutes, at twenty minutes
to eleven one day last August, and afterwards
ascertaining that the magnets at Kew were disturbed and deviated to the utmost during those
very ten minutes, after which they resumed their
usual daily course.

November 22. Talk with Mr. Huxley concerning C. Darwin's book, ' Origin of Species.' H.
thinks the theory admirable and a great step
towards the truth, far beyond Lamarck, inasmuch
as it assigns to nature a power of selection, and
so accounts for the appearance and modification
of species by secondary causes.   The book will be
attacked by naturalists, and by many religious
people, but H. thinks the time has come when
men of science should refuse to be snubbed by
parsons, and though there are persons of
enlightened   views,   such   as   Baden-Powell,
Maurice, Kingsley, Jowett, he cannot respect
their views, since while putting them forth with
one hand, they are ready to sign the Thirty-nine
Articles with the other. General Sabine, speaking
to me of the same book a week ago, said that
Darwin would prove to be an instance of a man
of first-rate science sacrificing all his reputation

to an idea; an idea as illusory as that of
Lamarck's, wherein it seems to me the General is
mistaken. Lockhart Clarke told me of his having
dined with G. H. Lewes, Miss Evans and Herbert
Spencer at Wandsworth, and enjoyed an evening
of truly philosophical talk. That Miss E. com-
bines a masculine intellect with feminine grace
and liveliness, and in a question of social science
defended her position with great skill and ability.
One of her reproaches to society was its indifference to, or want of high, moral rectitude. L. C.
describes Spencer as a remarkably good-looking
man, with thoughtful forehead, and as good a
talker as a writer.

December 8. Talk with Mr. Darwin, concerning his ' Origin of Species.' He is quite prepared
for all the criticism and censure which the book
will bring upon him : and with regard to those
naturalists who accept his conclusions as to
several prototypes, but not that of *one* primordial form, he believes that as they think on the
subject they must, in time, come to the same
result as he does. With regard to Owen's
dogmatic objection : that we have no fossils
exemplifying the transmutation, they may be yet
to be discovered, or may have been of a nature
to leave no trace. Is there not the lepidociren,

half fish and half reptile, and does not Owen show that the camelopard and the pig are but varieties of the same type; the same elements of structure prevailing in both?

Half-an-hour afterwards had a talk with Mr. Babbage on the same subject. He is quite convinced that the development theory is the true one; that an intermeddling series of creations are not the work of an All-Powerful, but that an endless succession of incidents and events planned to grow out of one primordial form *is* the work of an All-Powerful. He told me of an illustration he once showed to Dr. Lloyd, Dr. Robinson and Malthus; that his calculating machine, if set going with a series, might be arranged to go on for a million and two, and then change to another; that by inserting a slide he could prearrange that the change should take place at that number, or he could step in and insert after the million and two were complete. Which, he asked them, would prove the ablest mechanician, the first or the second record condition? Again, with the analytical engine, a certain series could be produced for ever; "but, Lloyd, you shall say how many ages that shall last: you, Robinson, say what shall succeed it: you, Malthus, decide whether you will have but one example of this

succession, and thus a miracle or a long series."
The force of the argument struck the party.
Matthews accepted it eagerly, Lloyd with anima-
tion, Robinson mused a moment, a cloud passed
over his face, but it cleared off and he admitted
the possibility of the development theory from
one primary law. The cloud, says Mr. B., was
simply a thought, "how will this affect me in
my profession," the Doctor having the organ of
shovel-hattedness.

Mr. B. thinks he will write another Bridgewater treatise and speak plainly what he thinks
on these questions, regardless of consequences.
Of revelation he says, that man's reason is
his revelation, and says, "what evidence would
convince any individual that he himself had
received a revelation? how could he satisfy
his own mind? by what evidence shall he convince a second or third party that he has been
the subject of a revelation?" Cannot imagine
what is the plan of the Deity in placing man
on the earth and endowing him with reason.
Believes that knowledge *per se* makes men
better.

November 8. At our evening meeting Dr.
Addison's paper 'On the effects of sherry wine on
the human blood-corpuscle' was read. The Dr.

generalises—thinks the serum or plasma of the blood affected only in inflammatory disease : the corpuscle in other diseases. Mr. Busk replied that the microscopic phenomena on which the theory was built were simply coagulations of fibrin. Mr. Turnbull, a visitor, asked me whether Addison was present as a spectator : and suggested that as the paper stated that John Amor's sherry alone would answer for the experiments, it was a case of *Omnia vincit amor.* This Turnbull is the man appointed to the State Paper Office, about whom there has been much discussion and censure in the papers because of his being a papist. He says he can bear all their buffets, and feel none the worse. Bruce, he says, is Unitarian; Mrs. Green is Methodist; and —— is a Jew. Among the papers, he has lately found a letter addressed to the Grand Master of Rhodes, telling that Antichrist has been born in Syria—that it has claws, talons, a fierce look, and diabolic expression, and was as lively and strong at the age of six weeks as an ordinary child of six months. When the Master of the Rolls [Romilly] read it he exclaimed with a chuckle, " I'll send it to Shaftesbury."

December 20. Talk with Sir C. Lyell. He says he had not given the opponents of Darwin's

book credit for sufficient discernment to see that
if they grant his argument so far as he puts it, it
must certainly lead to the conclusion that man
is not a special creation. But they do see it, and
were not a little troubled. "It is a grand book,"
he says, "one that shows scientific men are not
afraid of the parsons." He mentioned the
preface to Dr. Hooker's 'Botany of Australia,'
as an able and convincing argument from the
same point of view as regards botany.

K

1860.

February 25. Dr. Sharpey, while writing the Council minutes, talked with me of sundry matters. He said on the lunch table of the Athenæum there is at times a boar's head. Hart the artist, a Jew, stood one day looking at the head. Landseer, coming in with a friend, whispered, "Do you know what Hart is thinking about? Almost thou persuadest me to be a Christian." A man once eating roast pork with great relish wished he were a Jew. "Why?" "Because then I should have the pleasure of the sin as well as of the eating."

March 16. To Chelsea to tea with Alexander Gilchrist at 8 p.m. Went next door to call on Thomas Carlyle. Found him in his long dressing-gown, and his wife in the drawing-room represented by Tait in his "Chelsea interior." Tea on table, opposite the fire the large picture, a gift of Lord Ashburton's, showing the infant king of Prussia in the garden, as engraved in a frontispiece to the 'Life of Frederick.' Mr. Venables, a writer in 'Saturday Review' and 'MacMillan's Magazine,' had previously arrived. I was struck by Thomas' youthful and fresh-coloured appearance, remarkable in a man of sixty-three. Talk having arisen about an abortive search for engravings of Lord Ashburton's seats, he took down a volume of 'Beauties of England and Wales,' and read part of the description of Highclere, which becomes so rapturous over the charming scenery that we all laughed, and he shut the book, saying it was too pathetic, but we could have more of it if our feelings were not sufficiently touched. His face was a good study while he read, wearing no spectacles, shadows thrown on one cheek, bushy eyebrows, moustache and beard masking the massive underlip. A cheerful tone of voice, the more reason by its occasional manifestations of native Scottish. Then Kingsley was announced: entered in hat and overcoat and with a hoarse voice from a cold: in mourning for his lately deceased father. Enquiries as to health led to his telling us that for three years to come he was forbidden to write or undertake severe intellectual work, he having had warning and acute pains in the head, accompanied by great heat. Thenceforth the talk was mostly monopolised by him and Carlyle, both approving the volunteer movement.

C. " 'Tis a good thing for all that number of men to get themselves washed and cleaned and used to punctual habits."

K. " 'Tis one of the best signs of the times, full of hopefulness. Only think of the good that will be done by all those men having something which they must obey. You shall do this, you shan't do that, and made to feel that it is so, and not otherwise. Yes, if only to learn that they must obey, it is a great good."

Then C. spoke of the Eskdale and Annandale Militia coming into his native place when he was a boy; wild, big unkempt fellows, who however were dressed in uniform, and after some weeks of drill were good soldiers, and the whole regiment fired their volley as one man. He spoke of being much at Aldershot and Sandhurst, where he saw militia in no wise inferior to the line : of the rising officers at Sandhurst, men of twenty-five or twenty-six, thorough gentlemen and soldiers, ready for anything, pushing on the old men of the old school above, solely by reason of the elevating tendency of their own attainments. He had no misgivings as to generals. There was Mansfield, the man of most promise in the army;

but what he feared for was for the navy. Where
were old admirals? I mentioned Sherard
Osborne, McClintock, Keppel, Shadwell. "All
very well, but we want twenty."

Then C. "I am lost in utter astonishment
at our army; a man is put at the head of
50,000 men, and sent away in charge of the
honour of England, knowing no more of his
duty than if he were a wooden pole crowned
by a cocked hat; true men *had* turned up.
Wellington did; but W. was not to be compared to Marlboro': he was a practical man,
one who understood that causes produce effects,
that if a barrel were not well hooped it would
not hold water."

K. supported the Duke, instancing Torres-
Vedras as a master stroke, that when appointed
to the command, he said, "Napoleon had con-
quered because every one was afraid of him, that
he was not afraid, and he showed that after
all Napoleon was not God Almighty." And concerning the Crimean campaign, "that whatever
were our mistakes and losses, the French were
tenfold of ours."

Then some general talk on Macaulay's falsi-
fication of history; then literature generally,
and K., "How long will this jackassery, this flood

of books written by people who have nothing to say, continue? Look at Dickens, a man who might have been a Defoe if he would but have restrained his pen, who has degenerated ever since 'Nickleby,' whose Christmas stories are gloomy and depressing."

"What is the reason?" I asked. " Ignorance; he is one of the most ignorant of modern writers."

C. "I find the humour of his 'Pickwick' very melancholy. As for Defoe, he would have been a greater man, but he was such an incontinent fellow; always write, write, write, on some petty city matters. But he had wonderful power of imagination, making you feel he had seen everything he described."

Then K. spoke of memoirs of Col. Carleton, and fighting Peterborough, who would have been a great man had he not died young. C. replied that he sometimes passed Peterborough's house at Walham Green, when taking his rides. Then sermons were talked of and the strictures on books applied to them. "I hate the sound of my own voice," said K., "especially if I have to speak beyond a quarter of an hour. 'Tis a torture to me."

Then I: "Then every Sunday is to you a martyrdom."

" It is, and judge of my feelings when I am obliged to listen to somebody else's sermon for thirty-five minutes. Think of 15,000 clergymen having to stand up Sunday after Sunday with nothing to say. Ah! the Reformation has much to answer for." Turning to C. "You and your Puritans have much to answer for. Those men first started the notion that the way to heaven was by infinite jaw: and see what infinite jaw has brought us to."

" Ay," said C. " 'Tis wonderful how men will go on talking with nothing to say."

Then comic literature. K. "We are told it does harm. I have looked at it, my boy at Harrow reads it, but it is a decided improvement on 'Tom and Jerry' which I, and boys of my time, read when at school. Dickens' 'History of England' was sent me for my school at Eversley. I read it, and instead of giving it to my clods, put it into my fire." 'Soapy Sponge' had gems of wit. ' 'Handley Cross' some first-rate passages, but Surtees the author was degenerating. "Yes, he has a celerity in falling," said C. At ten K. rose to go: Venables told the story of the Welch preacher and his tale about Noë; how that Noë worked at the Ark driving in nails, plump, plump, plump. The haythen came and said, "Noë, there's good hunting in the woods here, hares and foxes, leave your work and come and hunt;" but Noë kept on hammering, plump, plump, plump. The haythen came again, "Noë, there's good beer at the Red Lion, leave your work and come and drink;" but Noë kept on hammering, plump, plump, plump. And then the rain came, and the flood lifted up the Ark, and carried Noë away, and left the haythen all screaming and squabbling in the water.

May 18. Left a large portion of my manuscript of 'All Round the Wrekin' with Chapman.

May 31. Finished my book, and with a sense of thankfulness, 527 octavo pages.

June 23. To see the review in Hyde Park. Got a good place in the front row of a gallery; an animated sight to see the large field fenced by thousands of men and women, and a sprinkling of red coats. The brigades marching to their positions, bands playing, one dense mass after another. Then the salute, flash after flash, and arrival of the Queen. Such acclamations! Then the *cortège* preceded by the Guards with snowy plumage, followed by a similar escort. Then the marching past, company after company, the scarlet-coated Huntingdons, the sober-suited grey, the sober but earnest sentiment of England, the bands of gay uniform, artisans, clerks, traders, lawyers, brokers, not to defy but defend; to look and see how they still came was impressive. Then the whole 20,000 in clumps advanced in one line and saluted, a most impressive spectacle. Where the ground curved could only see top of caps. Acclamations, caps on rifles, Queen departs, crowd break in, all good humour, fraternise with men from Kent, Durham, Lancaster, a sight at which an Englishman may feel proud.

August 21. Called on Dr. Sharpey on my way home: stayed to tea. We talked about sundry Royal Society matters, and the probable successor to Sir B. Brodie, whose dimness of sight may necessitate his resignation. The likeliest seemed to be Lord Harrowby, Duke of Argyle, Lord Brougham, Lord Chief Baron, Sir Philip Egerton. Serious difficulties appear in the way of a biennial Presidency.

September 4. Dr. Stenhouse called at Burlington House, having just returned from a visit to Scotland. . . . He saw the Paraffin Works at Bathgate. The coal from which the oil is distilled is a species of cannel-coal. The estate on which it is dug once belonged to the Honey-

mans, one of whom was Lord of Session . . .
now it belongs to a Gillespie, who let it at a
royalty of sixpence a ton on the coal dug.
When he heard that the Company were getting
fifty-six shillings, he commenced an action against
them on the plea that the mineral was *not* coal,
that he had agreed for coal.   He was defeated,
but raised so much litigation that the Company
were obliged to pacify him with £30,000.
Their return last year for income tax was
£100,000.   They make no gas, nothing but
paraffin or candles, and as the oil costs them
but a shilling a gallon, they might sell it at a
smaller price than the present.   The patent
will expire in four years, and then the oil will
be cheap.   Any kind of petroleum will yield
paraffin : the Rangoon tar, Trinidad pitch,
asphalte.   There is good asphalte in Cuba; if a
big pot were set up and one portion burnt to
reduce the other, there is a fortune to be made
by whomsoever will undertake it.

September 17. General Sabine returned from
a visit to Norfolk: tells me he met Elwyn, late
editor of 'Quarterly': a clever man and agreeable in society.   He gave up the editorship
because of the great labour.   His successor is
a Macpherson, who has lived some years in

India. He told the General that he (Elwyn) was once at a dinner in London at which Sir R. Murchison presided. Sir R. presented himself wearing all his orders, and his after-dinner speech was to take each order in turn in hand, give a brief history of the monarch whose effigy was stamped thereon, and then, " And His Majesty honoured me by the gift of this order."

Elwyn said great labour was involved in reconstructing articles, and in eradicating the personalities indulged in by the authors.

October 17. General Sabine told me that Sir B. Brodie is now quite blind—that he will resign the Presidency in January—that he (Gen. S.) will deliver the anniversary address. Many Fellows are urgent that Sir B. B. should resign —many dislike the General, and Dr. Stenhouse intends to raise a discussion concerning the increase of Secretaries' salaries.

December 17. To Mudie's at 9 p.m.; a gathering to celebrate the opening of his new library hall. Had talk with R. Chambers, who has just returned from America with his brother David, who tells me they think of going into the general publishing business on conditions very favourable to authors, allowing them an annual percentage on the sales. He introduced me to Captain and Mrs. Mayne Reid. I saw Sala, S. C. Hall and wife, Monckton Milnes, Washington Wilkes, and sundry celebrities whom I had not seen before. It was a success for Mudie.

CHAPTER III

WALTER WHITE AND LORD TENNYSON

1845.

October 15. Alfred Tennyson, it is reported, has a pension of £200 per annum.

1846.

March 3. Alfred Tennyson obtained his pension through Mr. Hallam's influence. Mr. H.'s son was about to marry T.'s sister when he died at Vienna. Mr. H. now pays to Miss Tennyson the £300 annually which he used to allow to his son.

1849.

December 12. Saw to-day Charles Tennyson
Turner, Alfred Tennyson's brother. He was
married long since to Mrs. Weld's sister. A
separation took place, and within the past few
months after thirteen years of disunion they are
reunited. The suffix of Turner was adopted in
consequence of a bequest of £500 a year with
that proviso. The two have come to town to
purchase a carriage and other matters. He,
Charles, very much resembles the portraits
of his brother Alfred, an earnest thoughtful
countenance. Alfred is still lounging at Chel-
tenham. A domestic thief has lately stolen a
quantity of his loose MS. poems (as I hear) of
a somewhat elegiac character. Also a new edition (the third) of 'The Princess.' He loves the
mild winters of Cornwall, and rambles among
the wild scenery in stormy weather. Recently
being there he wished to make enquiry respecting
some ancient monuments, and addressed himself
to a sexton, who referred him to the clergyman.

The latter answered the queries, and introduced the subject of Tennyson's poems, and then added, " Dear me, Sir, the portraits I have seen of the poet resemble you exactly ; are you Mr. Tennyson ? " On receiving an affirmative answer the minister flew to the bell exclaiming, " Emma, Emma, come down, Mr. Tennyson is in the house."

December 29. Weld home from his Christmas visit to Mr. Bawnsley of Shiplake. Tennyson, he reports, is in Chambers, 58 Lincoln's Inn Fields, engaged in the sequel and completion of ' The Princess.' He (Weld) read to me " The Bugle Song," which forms part of the new poem, and very beautiful it is. " The soft light shakes, across the lakes," then the description of " echo, thinner, clearer, in the purple glens dying, dying, dying—the horns of elf-land," etc. etc.

1850.

April 12. Weld read to me three new poems by Tennyson, part of a volume about to appear. They will be chiefly of an elegiac character, on the death of young Hallam.

June 13. Just as the meeting ended Mr. Weld returned, he had been to Shiplake to be present at the marriage of Alfred Tennyson with his sister-in-law Miss Sellwood. The happy pair went to Pangbourne for their first stage.

June 15. 'The Athenæum' has a notice of Tennyson's new book 'In Memoriam.' The Literary Gazette thinks it pretty good, and attributes it to a lady.

August 5. Went to London Library to get books for A. Tennyson.

October 23. Tuesday. Was introduced to Alfred Tennyson. He realised my idea of the poet, the lines of his face more decided and deeper cut than in the published portraits. He is tall, perhaps 6ft. high, has a sort of shuffle in his walk, wears spectacles, and speaks with a rather heavy and full voice. He thanked me for the little services I had rendered in the summer, and his wife promised me a copy of the book. Weld told me of his doings yesterday in house-hunting with the poet. The latter is difficult to please in the matter of a house, he must have all the upper part to himself, for a study and smoking-room, etc., and to avoid noise above his head and indoor privacy, as he is accustomed while composing to walk up and down his room loudly reciting the flowing thoughts.

November 14. Weld told me this morning that the Laureateship has been offered to Tennyson.

December 15. Alfred Tennyson in our office for an hour to-day. Gave me his autograph in the copy of 'In Memoriam' which he lately presented to me. Called his attention to the paper on Arthur and the Round Table in the Berlin Transactions.

1852.

September 29. Tennyson called in the afternoon to show a friend the Library, etc. On seeing the bust of Banks, he related that Sir Joseph was once dining with his father and said, "Dr. Tennyson, I have tasted almost everything in my life, animal and vegetable, but there was only one thing that turned my stomach, and that was a boiled bug."

October 16. Tennyson came to the Library to-day. After a time he said, "I must have a pipe." Mr. Weld replied that he should either go and smoke up the chimney in the back library or on the roof. He chose the latter, and I went to show him how to thrust his huge length through the window. In a quarter hour he came down again greatly refreshed. During a conversation on French affairs on the day of the christening of his child, he broke in with his deep sonorous voice, "By the holy living God, France is in a loathsome state."

November 5. Tennyson in town from Seaford,

where he is sojourning for a time with wife and child. He has finished his Wellington Ode and has come to town to sell it. Weld tells me it is majestic in its march and language, and contains a grand invective against Louis Napoleon and France which will be omitted. Parker came at noon; he offered £150 down for an impression of 10,000 to sell at one shilling. Tennyson will accept this, if Moxon, to whom he went forthwith, is less liberal in his offer.

1853.

March 6. To Twickenham to pass the day by invitation at the Poet Laureate's; arrived about 3 p.m. Walk with Mr. Sellwood by the riverside before dinner: he showed me the avenue in which Sir W. Scott makes the interview take place between Jeanie Deans and the Queen.

April 7. Was much gratified this evening on opening a small parcel addressed to me, to find it was the eighth edition of Tennyson's Poems from the Laureate himself, and bearing his autograph.

1854.

May 21. Went this afternoon by invitation
to meet Tennyson and Weld at the Zoological
Gardens. The ground looked beautiful—flowery
and spring-like — the red May in perfection.
We walked about looking at birds, beasts, and
fishes; fell in with Gould who talked zoologic-
ally. The vivarium singularly interesting. From
thence we walked down to Somerset House to
dinner. The Laureate was obliged to go away
at eight o'clock to a party near Portman Square,
and at eleven he is to go to a smoking club
which never meets but an hour before midnight,
and of which he and some great man of Darm-
stadt are the only two honorary members.

1858.

June 15. Tennyson has been in town a week, staying at Burlington House with the Welds. Have had speech with him sundry times. He is much alarmed about a French invasion. To-day he read another batch of poems to some friends, of whom one was John Foster. In these poems there is another chapter of ' Morte d'Arthur,' and of such touching words that J. F. shed tears.

November 11. Tennyson in town discussing new terms for the publication of his book with Moxon's trustees, Spalding and Evans. The late Moxon always took one-third of the profits and five per cent. besides on the gross amount of sales. Now the terms are to be ten per cent. only on the books sold. Had the Laureate not been of a yielding disposition the change would have been made long ago.

1859.

March 24. Tennyson in town. He read to
me one of the chapters of his Legends of Arthur
. . . a grand musical intonation in his deep
sonorous voice. We stood by the mantelpiece
in my office, and he read on one hundred and
forty pages, till the story was finished. 'Tis
admirably told, the contrast between Merlin's
and the harlot's nature well sustained, and the
way in which he yields at last is most skilfully
conceived. There is however that in it which
will shock the "unco guid" folks. Spoke my
mind freely and advised him to publish. Dined
with him at Weld's, Barrett and P—— of the
party. The latter an incessant babbler.

November 11. Weld tells me that Macmillan
has been on a visit to Faringford and has given
Tennyson £250 for leave to print a poem 'The
Three Clerks' in the new magazine. Seems to
me a wild speculation.

1860.

August 10. Dined with Mr. and Mrs. Tennyson at Burlington House. Palgrave and Woolner of the party; Spedding came in at tea-time. The Laureate talked of going to the Levant, to West Indies, Cornwall or Brittany.

August 14. Talk with Tennyson concerning Dartmoor, Cornwall, and the Scilly Isles. He thinks of writing something more about King Arthur, urged thereto by Mr. Gladstone and others. If he does there will be a chasing and marching of Arthur and Sir Modred from Tintagel down to Lyonesse, the now submerged region beyond Land's End. He commonly composes while smoking, and keeps the lines long in his head before he writes them down; dislikes the labour of writing and so loses many thoughts by delay. He once had three hundred lines in mind concerning his Lancelot and his quest for the Sangrail, and lost them all through leaving them too long unwritten. Does not remember if he has written more in one place than another;

writes wherever he may happen to be. ' Locksley
Hall' was written at High Beach in an old house
which has since been pulled down. He admires
the Vale of Thames between Maidenhead and
Streatley, and now departs for Oxford preparatory
to journeyings in Devon and Cornwall.

The place where Tennyson composed 'Locksley
Hall' seems to have been an object of much
interest to Walter White, and we find that the
poet in 1870 told his friend that the house used
to be called Beech Hill, and was pulled down,
and that Richard Austin, son of the Serjeant,
built another on the site. Tennyson added that
much of the composition of the poem in question
was "seen and heard" (referring to the poet's
habit of reciting his work before recording it) at
2 Mitre Court Buildings, Temple. Being pressed
for a personal opinion, Tennyson characteristic-
ally added, that this kind of literary gossip was
not interesting to him when related of others;
and was not particularly grateful to him when
printed about himself. The matter is perhaps
one of much greater importance to the reader
than to the writer of poetry.

1861.

November 16. Chapman junior came to me in the office to say that Tennyson was in their counting-house and wished to see me. I went across and found him in conference with the two partners, having told them that his engagement with Moxon's was not likely to be broken. He left presently, took my arm, and we walked down Piccadilly together. He said he could not yet muster courage to come to Burlington House after what had taken place, that he will most likely when next he comes to town, meanwhile hopes that I will go and see him. Told me of his visit to France, his disgust at the bad food and stinks of the hotels and boarding-houses. Asked where I had spent my holiday, and promised to give me a little information as to where his poems are written.

1862.

May 6. Tennyson and Woolner called on me while I was at lunch and partook of ale and bread-and-butter. We told a few stories and had some talk. The poet has written a poem embodying a description of a tropical island, and wishes to see a good view of an Isle to verify his description. The publication of his ode in ' The Times' was surreptitious, some one at the Exhibition having copied it from the music, and seeing *p.* for *piano* before " art," wrote " part divine," which had to be corrected by the author in a letter to the paper. He complains of the view of the sea being built out from him at Faringford, and says he shall depart for a time at the end of this month to get out of the way of Cockneys.

1864.

February 13. Tennyson called on me this afternoon. I gave him two sorts of Hungarian wine to taste. He liked the Oedenberg. I asked him about "Enoch the Fisherman" which I had seen mentioned in the 'Reader.' He answered that he had had a proof more than a year, could not yet make up his mind to publish. His friend Spedding liked it. He had been at Palgrave's the evening before; the Eastern Palgrave was there and talked of his travels, every one listening intently. Gladstone, Chancellor of the Exchequer, called in, and he listened with the others. Then the Laureate recited a ballad poem in the Lincolnshire dialect; an old farmer on his deathbed talks to his maid; will have his beer (yale); tells her what the Doctor and the Parson have been talking about. He stayed an hour, and on departing, invited me to breakfast with him and Woolner at Spedding's.

February 14. At 9.30 to Spedding's, 60 Lincoln Inn Fields. I was looking at the old engravings on the wall, when Tennyson entered and proposed a quarter of a mile's walk before breakfast, his usual practice, he said. So we walked up and down the north side of the Fields in the sunshine, he trying at times to outstrip me, but I told him I had learnt *tall* walking in America. As we returned to the house we saw some spilt milk on the pavement, and he would finger the congealed liquid to try whether it was really milk or chalk, and said he had once seen milk spilt on a doorstep evaporate and leave a layer of chalk! "We should be mobbed," he said, "if this were a week day." Soon after we remounted to Chambers, Spedding entered, a man with a Czechist's form of head, thoughtful-looking, and exact in his speech. Presently Woolner came and we sat down (Tennyson introduced me to Spedding as "the author of Northumberland "). "No, not exactly that, for 'Northumberland' was created before he went there." (It reminds me of 'The Heavens' by the author of 'The Earth and Sea,' one Mudie.) There was no remarkable talk. After breakfast Tennyson and Woolner retired to smoke. Spedding and I talked till 12.30. . . Then I went into the adjoining room to say good-bye to the smokers. Tennyson told a few good stories and invited me to Faringford.

March 25. At 9 a.m. by steamer to Yarmouth. Much tidal marsh land below the town. On right bank near mouth the site of the projected docks, about half-an-hour's trip to Yarmouth. Walked hence to Faringford. Came to Tennyson's at ten, had an agreeable welcome. Introduced to Mr. Lawrance, who was taking a crayon portrait of the Poet. Then talk with Allingham, a rhymer not unknown to fame. Then walk in plantations amid primroses and daffodils. At lunch introduced to Professor Jowett and Mr. Wilson, the tutor. Afternoon walk on the Downs. Dine at six.

Saturday. Football with the boys, and ramble with Allingham and the Laureate. In the evening I introduced the subject of pronunciation, with a view to learn what opinions were as to the subject. I contended that pho*to*graphy should be pho*tography* and the like. Sunday, to Freshwater Church with Professor Jowett and Hallam Tennyson. Afternoon, called on Henry Taylor, author of ' Philip van Artevelde,' and a walk to foot of cliffs by Watcomb Bay and on the Downs. After dinner, talk about Cosmology.

March 28. After breakfast the poet showed me the proofs of his forthcoming volume. I had only time to look at one or two little poems, and

to see that there was one on ' Boadicea,' and at
11 a.m. I departed, walked to Yarmouth, and
arrived in London at 6 p.m.

November 20. Going out at 1 p.m. met
Tennyson in the courtyard. Turned back, gave
him lunch and opened a bottle of Oedenberg for
which he asked. Told me he had sold more
than 40,000 of ' Enoch Arden' and cleared more
than £5000, that he had abusive letters from
people who blame him for accepting so much
profit. Talked of his project of sixpenny
numbers: said Pyne, Moxon's manager, expects
to sell 50,000 copies. If this expectation be
not disappointed the profit will be £10,000.
Talked about the encroachment of buildings
around Faringford, and the villas and hotel
that are to be built at Alum Bay, spoiling the
solitude of the Downs, and that the railway is
to pass on two sides of his land. I suggested
his buying an estate of heathland, ninety acres
including a hill, one of the Devil's jumps, at
Chart near Haslemere, which was in the market
for £1500. He seemed to like the notion but
started objections. He then proposed a walk,
and I proposed the Zoological Gardens. We
called for Woolner, who gave me to taste a
new kind of Spanish wine, Ampurdam. As we

walked about the Gardens, Tennyson said, as we looked at the white peacock sitting crouching and with loose feathers looking somewhat strange in the damp air, that people had written to ask him what he meant by " Now drops the milk-white peacock like a ghost," and Woolner broke out, " Why, don't they know that a mass of white always looks ghost-like in the dusk ? " We met Bence-Jones, to whom I introduced the Poet, and he (B. J.) took occasion to beg him to give a recitation at the Royal Institution.

November 21. Tennyson called to enquire about the estate at Haslemere.

December 22. He called again with Mr. Woolner and proposed to meet me at Hasle-mere with Mrs. Tennyson and go and look at the land, which, being heath and having a hill and a brook on it, has charms for both of them.

1865.

July 9. To breakfast at Woolner's, Tennyson, Spedding, and Mr. Pollock. We had talk about spelling, about a universal language, about Bell of Edinburgh's mode of writing sounds of any language so that any one ignorant of the language can pronounce the sounds. Then when pipes were lit the Laureate read us a poem. Another Northern Farmer, who had made money and enlarged his farm and is proud thereof; and mounted on his horse "property" expostulates with his son, who wants to marry the pretty but penniless daughter of the curate. After that Macmillan and Lawrence and Arthur Hughes and Fenn came in. The latter talked about getting Gustave Doré to illustrate an edition of the Poems.

December 7. Tennyson, Woolner, and Dr. Sharpey dined with me. We had merry stories and grave and cheerful talk. At 8.30 to the evening meeting, at which the Bard was admitted

and Prof. R. Grant of Glasgow. At the end of
Cayley's paper on 'Ischirnhausen's Transforma-
tion,' Tennyson and Woolner went back to the
dining-room for a smoke. Then at the close of
meeting to the Lower Library for tea and talk.
Attracted by the portrait of Copernicus ; and
then I showed them Galileo. The former they
thought imaginative and grand, the latter has
the painful look of one in mental sufferings.
Woolner told me the Chancellor of Exchequer
(Gladstone) is to dine at his house to-morrow,
as Tennyson says, " to meet the poor Poet."

December 16. Tennyson, Woolner, and Cresy
dined with me. The Bard spoke of a man who
persists in writing to him, addressing the letters
to *Miss* Tennyson, and reproaching him sorely
with having made £5000 by 'Enoch Arden.'
He said he often gets letters of enquiry as to the
meaning of passages of his poems, " as if," he
continued, " I could remember. I knew what I
meant when it was fermenting in my brain, but
how am I to tell now what I meant then ? " At
twenty minutes to twelve we found ourselves
talking of free will and necessity, and then the
party broke up.

M

1866.

October 20. To Haslemere .... to look at Singleton Lodge and heath land, a property which Tennyson looked at a month since.

November 5. Talk with Mrs. Gilchrist about the lands at which Tennyson is to look. At one o'clock to the station; met the Poet and Mrs. Tennyson and Payne. After lunch a walk to the Hill Copse, commanding a glorious view towards Little Hill, and all the diversified landscape between. Heavy misty rain. The Poet was pleased with the place and made up his mind to buy it. He would like the spring better were it a rush of water. The firwood on hilltop and expanse of wild heath to rear charmed him.

1867.

June 29. At 5 p.m. to Haslemere. Dinner on the lawn. The freshness an agreeable contrast to the heat of London. Sunday morning to Marley. Sat enjoying the prospect and reading Danish. After tea a walk to Greenhill on Blackdown, the estate of thirty-five acres bought by Tennyson, a charming spot—copse, ferns, brambles, slopes, fields, background of heath, and grand prospect to S. and E. We saw the sea twenty-five miles distant, looking over Littlehampton.

1868.

February 14. Called on Woolner, found him in his sanctum busy with his sketch of Palmerston. .... Woolner told me Tennyson was in the house, front room second floor. There I found the Bard smoking. He had been on visit to Cambridge, to the Vice-Chancellor, had stretched his limbs, he said, in that lodge which he regarded with awe as an undergraduate. The V.C. offered him the LL.D., but he declined it. Would not object to a diploma to that effect, but does object to the public ceremony. The Prince of Wales had enquired after him while travelling. " How should the Prince know I was in Cambridge ? " said the Poet. " Was it in the papers ? The papers tell so many lies about me, say so many things which I know to be lies, that I suspect the other things they say to be lies also." Through the Duchess of Sutherland the Queen offered him a baronetcy. He declined. " Can the Queen do nothing for you ? " · " Yes, if she would shake my two boys by the hand, it would help to keep them loyal." This the Queen afterwards did, and Tennyson dined at Osborne with Lady Augusta Stanley. The Queen drove him out. He told her Osborne would be a pretty place thirty years hence, meaning that at present it was too naked for want of large trees. He spoke of his new place at Blackdown, that he is expected to lay the first stone next month. Admitted that ultimately he would have to leave Faringford, so much is he annoyed by the building of houses there. I reminded him that his compensation would be in the large sum he would get for his estate, and in the amenities of his new place, and I reminded him once more of my suggestion to write a poem entitled ' The Beacon.'

April 18. Letter from Mrs. Tennyson, inviting me to the laying of the first stone of their new house on Greenhill, Haslemere, next Thursday, 23rd. Unfortunately a Council meeting here will prevent my going.

April 27. Mrs. Gilchrist informs me that the laying of the stone went off well. But a small party, Sir T. Simeon, his wife and daughter, the Bard, Mrs. G., Mr. and Mrs. Knowles, the Simmons'. The stone block, of native sandstone, had on one side Æ, and on the other, " Prosper the work of our hands upon us."

March 10. Macmillan called ; took me to his house near Tooting Common, with G. Bell (Bell and Daldy) to dine. The house, which is old, built at different times, is roomy, a good-sized library for smoking-room, stands in about three acres of ground where are some good trees and rare old elms. Mac is famed for his dinners, and afterwards, in the smoking-room, he gave us a treat, the reading of Tennyson's new poem, ' Lucretius.' We all agreed it was the grandest thing the Bard had yet written.

1869.

April 25. To dine at Knowles'. Tennyson and Mrs. T. there, and Mrs. Knowles. Bard and wife in town to buy furniture for their new house at Haslemere. Had a pleasant evening.

July 6. With Macmillan. Home to his house at Balham to dine, an impromptu visit. Talked about a scientific periodical which Lockyer is to edit for him, about Tennyson's transfer of agency to Strahan. He, Mac, offered the Laureate £3000 a year. Then, while smoking, he (Macmillan) talked of his father and mother, his early days and struggles, surgeon's assistant, sailor before the mast, teacher and school-master, and last bookseller.

1871.

July 1. To Haslemere . . . . to Tennyson's new house at Aldworth, Blackdown : a palatial-looking

house on a small scale : large mullioned windows,
gable-lattice finials, pointed arched porch all of
stone, where the terrace dips a handsome balus-
trade, steps and vases, openings made in the
copse, and winding paths, a glorious prospect
which I had seen before. Met Farrar, F.R.S.,
Master of Marlboro' College, and his wife ; she
a very bright-looking lady. Hallam Tennyson,
the son, grown as tall as his father. Touching
to see how attentive he is to his mother. In
the evening the Bard read aloud 'Geneviève,'
and the introduction to ' Morte d'Arthur.'

1878.

April 8. To 14 Eaton Square, saw Mrs. Tennyson reclining on a sofa. Talk for ten minutes,
then came Knowles and Mr. and Mrs. Lewes.
While she bent low to talk to Mrs. T. Knowles
and I, and Lewes gossiped about Mallock and his
article "Positivism on an Island," in the ' Con-
temporary.' They pronounced it poor. About his
being about to become Romanist, and to marry a
Spencerite, about the state of mind that leads
people to Rome. Of Clifford and his illness, and

going abroad perhaps to die, of his answer to
newspaper report that he too was a pervert, viz.
that his M.D. had certified he was ill, but 'twas
not mental derangement, and he gave flat con-
tradiction.   Tennyson and his son Hallam came
in, and Mrs. Lushington, and we all went down
to luncheon.   Pleasant chat.   Mrs. Lushington,
who lives at Bath Hotel, promised to visit Burlington House and take luncheon with me.   Sub
sequently Mr. T. said he was going back to
Faringford, being tired of London, and the
attentions of King and Co.   Lionel and his bride
to return from Gibraltar by steamer.

CHAPTER IV

ASSISTANT SECRETARY OF THE ROYAL SOCIETY

1861.

January 30. Dined yesterday at Alex. Gilchrist's. Met Lowe the editor, and Crockford, manager of the 'Critic.' Lowe says that Urquhart, who was grey, has completely recovered the flaxen hue of his hair by use of the Turkish bath.

February 1. E. Chapman told me an anecdote of Sir Bulwer Lytton which he heard from Maclise. He, Maclise, once painted Bulwer's portrait full length : the author was fidgety concerning his feet—would have them made very small. Just before the finish Maclise painted them the natural size. Bulwer came to see, would have them diminished. "Well," said the painter to Chapman, "after the pictures are hung we are allowed two or three days to work on them, and I went and painted the feet in of their proper size." A half-length of this portrait was engraved for a book published by Chapman. Bulwer was fidgety over that also. Now this, now that was not right, and " the complexion of the hand " was especially to be cared for.

March 21. General Sabine . . . spoke of taking the place about to be vacated (on staff of the R. S.), and of a Mr. Wilson, who he thinks may be suitable. As usual his words and manner were very kind : the information however excites me, and I must be watchful and remember, " Let him that thinketh he standeth, take heed lest he fall."

March 25. Called on Dr. Sharpey, gave him my 'Working Man's Recollections' to read, and told him of my domestic circumstances. His answer was encouraging. He advised me to repeat the narrative to General Sabine. This I did on the 27th, when the General came to Burlington House. He listened attentively; said the tale did me honour.

April 11. An anxious period for my chiefs and me ended to-day by the acceptance of Weld's resignation. The President and officers are a committee to recommend as to a successor.

April 19. Weld for the first time spoke to me of the matter, thus opening the way to ask advice as to the steps to be taken for getting another post.

May 2. Our Council sat to-day. Dr. Sharpey came out and said, " Well, I come to tell you you are appointed assistant secretary, and unanimously." This was followed by such kind and hearty congratulations from Mr. Spottiswoode, Mr. Huxley, Lord Wrottesley, and others, as filled me with grateful emotions. And so begins for me a new career of responsibility. I trust, with God's help, to heed my good mother's advice to "use sound wisdom and discretion."

May 11. A very busy week with me in preparation for our President's *soirée*, and especially busy was this last day, rendered more so by the unmitigated rain and raw temperature. However, by 6 p.m. arrangements were so far complete that I could go and dine, returned, gave a few finishing touches, shut myself up in an archives room and dressed, and was ready to receive General Sabine at 8.30. A few minutes before the first guest arrived, and from that time the cheerful gathering grew. Satisfaction with the arrangements was general, and I got so many

congratulations on my promotion as to make the
time one almost of ovation for me. It was
Sunday morning before the last of the party
dispersed, and very tired did I feel when the cab
delivered me at my residence at ten minutes past
one.

June 3. Set the cleaners to work upon the
apartments in Burlington House which henceforth will be my residence.

June 16. My first Sunday in the new home.
Quiet and agreeable. Pleasant to walk on the
terrace, hear the rustle of leaves, and chirp
of sparrows. Like the place better than I
expected.

July 19. To Lord Ashburton's *soirée* at 9.30.
The house and paintings more interesting than
the curiosities collected. The presence of ladies
imparts a charm. Saw Lord Eversley the late
Speaker, and Lord Elgin. Had a long talk
with Thomas Carlyle, concerning his travels
in Bohemia, his dislike of the Czechs, and the
way in which a Czechist landlord cheated him,
because the bargain for a wagon had not
been put into writing. (Haben Sie etwas Schrift-
lich.) He said a good walk and book might
be made out of Luther's land and special
haunts ; that the Wartburg is one of the most

curious old castles he ever saw. I said I had often wished that he had taken up Luther to write about instead of the Great Frederick. He thought that enough had been written about Dr. Martin Luther.

1861.

September 17. Dr. Hooker called to ascertain
the date of his appointment on the Committee
which recommended a course of observation by
the Schlagintweit expedition to India. This is
in consequence of the review of Schlagintweits'
book in 'The Athenæum,' and the subsequent
explanation of Sir Roderick Murchison and
General Sabine. He blames the Indian authori-
ties severely for their passing by such men as
the Stracheys and Thomson in favour of Germans
who are less capable. That the Schlagintweits'
appointment was a flagrant job, but at whose
instigation he cannot tell. That Col. Sykes
told him while the brothers were in India and
when their work was but half done that they
had spent £20,000. That they wanted him to
describe their plants—that he offered to do it
as a public duty and to enlarge his knowledge
of botany, though not especially for their book.
That they refused, telling him a German botanist
would do it; but that the botanist refused owing

to the dislike he, in common with other servants, bears them. That they got a good swag out of the sum alloted to them, and bought a barony and estate near Munich. That Thomson spent all his pay in making collections, and then was refused leave to publish the results—that while the Schlagintweits were encouraged, the collection made by Wallich, Thomson and Stracheys were rotting in the vaults of the India House; wagon-loads, which cost £40,000 in collecting.

September 17. He says that the work at Kew increases more and more, is becoming more and more the centre of reference for botanical science. That his father as a boy had a longing to be King of Kew, that he gave up a good appointment at Glasgow for his cherished notion, and took Kew with £150 a year only.

October 19. Mr. Wheatstone says Dr. Roget told him that he (Dr. R.) was one of the prime movers in the question of having Lord Brougham as President of the Royal Society; that there are many reasons why his lordship should be preferred to General Sabine, ignoring the fact that by his sneer at Young's Undulatory theory in the 'Edinburgh Review,' he disgusted Young, who quitted the subject, and Fresnel took it up, and the honour of the discovery went to France; that there is nothing really new in his Lordship's optical papers, that the results are all calculable, and that the Académie reads his papers but does not print them.

November 5. Dr. Sharpey came to the office this morning and told me further with some excitement about Mr. Graham's changeableness in the matter of the Presidency and of the Treasurership : after consenting to undertake each post in turn he declined both.

I hear that Mr. Graham wished he had accepted the Presidency, and that Mr. Faraday says if it were offered to him now he would accept it.

November 16. Dr. T. J. Gray says his wife has heard from Miss Roget that there will be no organised opposition to General Sabine's election as President : that Lord Brougham said to a deputation, who waited on him, that he would have stood had the Council invited him, but that he will not stand in opposition.

1861.

Professor Graham tells me he worked six years at his liquid diffusion before he mentioned it, and made all the experiments three times, whereby he became quite certain of his facts : that the more complete a man's investigations are, the shorter will be his account of them : incomplete results appear to involve long statements. He finds that leather can be tanned by silicic acid, but does not know yet whether the leather may not be converted into flint.

November 30. Anniversary Meeting went off satisfactorily. Sir B. Brodie delivered the address extempore ; the meeting listened with respectful emotion. Notwithstanding threats of electing Lord Brougham as his successor, there were 78 votes for General Sabine out of 84. The Anniversary dinner was held at St. James' Hall ; about 65 present. Dr. Carpenter and Professor Sylvester, the medal man, made a speech, also Sir Roderick Murchison, Dr. Robinson, Sir Charles Eastlake, and the President elect.

N

On my return home at 10 p.m. I was painfully astounded by news of the death of my friend Alex. Gilchrist. Took a cab, and forthwith drove to Chelsea; there heard from the nurse that he had died at four this morning of scarlet fever after five days' illness.

1862.

February 2. To Chelsea, my first visit to Mrs. Gilchrist since poor Alex.'s funeral, and her return from Essex. . . . She read me a very kind and touching letter which she had received from Rossetti, in which, besides sympathy, he offered to write a biographical notice of Alex., to be prefixed to the 'Life of Blake,' provided Macmillan's consent, and to try to produce a likeness of our lamented friend.

April 5. Walked out to E. W. Cooke's, The Ferns, Kensington, his exhibition pictures being on view, two of Venice, Cartagena harbour, Alicante, Bay of Tangiers : the sunset views of Cartagena very beautiful. The whole house is a miracle of art, with multiplicity of pictures, sketches, ornamental art, ethnological collections, botanical and natural history specimens, and one fernery ; this last is enchanting.

May 8. Our meeting to-night well attended, a good discussion on the Appendix to Mallet's Holyhead earthquake experiments. Faraday,

Tyndall, Clerk Maxwell, Sorby, Fitzroy, W. Thomson, Mallet, Daubrée and the President spoke. Foreigners present : Forchhammer, Dove, Regnault, Delesse, Stos, Frémy, Captain Belavenitz. Stos said to me after the discussion, he was astonished at the dispassionateness of the speakers, that such a discussion in France would become violent and personal.

May 26. Called on Marcus Stone, found him copying Frith's picture of Railway Station—the picture for which the painter was paid 8000 guineas. I had a good look at it, and came to the conclusion 'twas not worth the money. The very best portion comprises the three heads of the sailor kissing his child, and the weeping wife and mother ; the expression is wonderfully life-like. Saw the picture taken down, locked in a case, and carted off to its exhibition-room in the Haymarket. In the copy which M. S. is making, all the figures were outlined in by an engraver's outliner who can do nothing else, he had £40 for his job. Marcus will have £300 for his, and has to make besides a second copy. Told him of the favourable notice in ' The Times ' of his picture, ' Artist's First Work ' now in the Exhibition.

June 7. To Westminster Hall and Houses of Parliament, to the Social Science *soirée*. An

interesting sight to see the well-dressed crowd
pacing through the corridor, seating themselves
in the House of Commons, gay cloaks and silken
skirts decking the rows of seats, some men
placing themselves in the Speaker's chair. Then
the Great Hall, decorated with shrubs, the
brilliant light, the long refreshment-table, the
band of the Coldstreams, an imposing spectacle
viewed from the däis. The music seemed to me
poor and inappropriate, fifty violins would have
been sweeter, and far more appropriate.

July 2. Had a large party (thirty-two), with
music and dancing in the evening; it was
experimental and satisfactory.

October 27. Went with General Sabine to look
at Argyll House, late in possession of the Aberdeen
family. It is a mean-looking house outside, and
somewhat dark within, but has capabilities of
space that may be turned to good account: a
meeting-room might be built in the garden, and
the library accommodated, all on the ground
floor. Of upper rooms and basement rooms
there are more than enough. The situation is
comparatively quiet.

October 30. General Sabine told me this day,
that Mr. Rainey (the agent who showed us
Argyll House) saw Mr. Cooper yesterday: that

Mr. C. says the Government have no plans with respect to Burlington House, and do not mean to adopt either Barry's or Smith's, and that there is no intention of disturbing the Royal Society's occupation.

November 20. An audience of two hundred came together to hear Professor Owen's paper on 'Archiopteryx Macrurus.' The composition of the subject was masterly and convincing as to the "ornithic character" of the fossil. The Duke of Argyll asked a question as to the wing-feathers : Mr. Gould thought they showed the bird to be one that had never flown. Mr. Owen answered, and pointed out the wing-bone, which marks a bird of sustained flight. He described the creature as having been left by the tide on the shore : the head, breastbone, some of the dorsal vertebræ, and the large wing-feathers had been carried off or floated away. Dr. Carpenter said a few judicious words as to the caution with which negative evidence should be rejected, seeing that though the Solenhofen quarries have been ransacked for fifty years, it is only now that this fossil has been brought to light.

December 17. I called Sir Charles Lyell's attention to Mr. Bowen's Paper on the "Development Theory" in the 'Mem. Amer.' He had read it, and said he did not think it convincing. Had heard of what Dr. Carpenter had said at the reading of Professor Owen's paper, and thought it highly judicious: as an illustration of fresh evidence turning up, Sir R. Murchison had told him that the skeleton of a crocodile had been found in Old Red, near Elgin, below the coal. A surprising fact, for hitherto it had been supposed that no animal of so high an order existed with that formation. "It will be disputed," said Sir Charles, "by the opposite party with as much obstinacy, as twenty years ago they disputed the discovery of reptiles in coal. When I was in Nova Scotia, a bone was given to me, which I showed to Agassiz; it showed signs of the coal in which it had been found, and he (Agassiz) said: 'If it were not for those I should say that it was the bone of a reptile. But it can't be that, and is a bone of one of the fishes.' I was convinced it was the bone of a reptile," went on Sir C., "and showed it to Wyman. He wrote and gave me a paper upon it, with leave to publish it, if on my showing the bone to Owen he pronounced it to be reptile. I showed it to him, and he at once said reptile. But notwithstanding this we shall have to fight the battle over again."

December 25. W. and I took our favourite walk through the Warren to Mapledurham avenue ; on our return (to Reading), found that dear father had taken a five-miles walk, and right cheerful did he appear after it.

1863.

February 4. This day my good old father died. He was in his eighty-fourth year, having been born in 1779.

February 10. At half-past three we buried our father, without parson, pall, or trappings. Friends and sons rose up to supplicate the Almighty, but for me it was a very trying occasion; only with great effort could I regain my composure.

October 5. Captain Maury called to-day. I took the opportunity to speak about Southern affairs, and the recent battle and defeat of Rosencranz. He thinks it favourable, but says that the circumstances of the South have never been so desperate as represented, and has no doubt of final separation and success. I said it seemed to me that one important move towards ending the war would be to burn Washington; he answered that the South have had that notion for some time, and only watch their opportunity to carry it out. I mentioned, in answer to his query as to what was new, the recommendation, by the British Association to Government, of gun-cotton as compared with gunpowder. He said that it was a very important question, because if gun-cotton could be used, sixteen pounds would do the work of one hundred pounds of gunpowder, seeing that two-thirds of powder is solid matter. And because of its utility in protecting channels. A ship, or ironclad, might enter the Mersey, and pass the forts, but that if the channel were mined they could be blown up. To effect this, the mine or receptacle must touch vessel's bottom ; difficult to buoy up powder on account of its weight, no difficulty with cotton.

This was the first time I had seen Maury ; he is about 5ft. 5in. in height, square built, broad forehead, walks slightly lame, one leg appearing to be slightly shorter than the other. Does not nasillate in speaking, and is quiet in manner.

1864.

February 22. Mr. Wheatstone tells me his arrangements with the Telegraph Company for London are satisfactorily completed. He put in £4000 ; they gave him £10,000 worth of shares, on which sums he gets 5 per cent. He has besides a claim to £17,000 worth of shares which he gave back to the Company, but which may be again passed to him as capital is paid. Besides, he has a profit of £5 on every pair of his instruments sold ; the price is £25 to £30 per pair, when first made it was £40. The sale amounts to ten pair per week. For instruments rented he receives £1 a year for five years. All this is automatic telegraph ; with this instrument, and a single wire, he can send 600 letters a minute from London to Dover. He is about to try if the same can be accomplished from London to Newcastle.

April 1. General Sabine called; he is in excellent spirits about the Gun-cotton Committee, which met for the third time in our Council Room yesterday. He says that five series of experiments were planned, that each will require five months to carry out upon each gun in the service. So if there are 100 different guns, 500 months will be required. In a jocular way he said he thought that the Committee, being now fairly started, would last as long as the Royal Society.

April 12. Returning about 5.30 p.m. from Kensington Gardens, I saw a little excitement among the onlookers at the throng of carriages, and stationed myself by the rail. Presently a carriage came up at walking pace, and in it sat Garibaldi, with the Duchess of Sutherland and two other persons. He looks short, sitting, and has nothing striking in his look; the chief expression seems subdued. He had on a low-crowned hat, and as he raised his arm I noticed the red tunic. I was glad to have set eyes on the man who pulled down the never sufficiently to be execrated Bourbons from the throne of Naples. One of the party with whom I went to Silvertown yesterday, told me he knew Garibaldi eight years ago at Montevideo; that he was a very lion in fight, and came out covered with blood; that his plunder was never for any purpose but to get food for his band, he was never suspected of taking a sixpence for himself. I have never had a better view of a remarkable person, for as his carriage passed he looked me full in my face, so that I had a complete view of his quiet-looking features.

September 20. Sir F. Pollock told me that on Friday next will be his eighty-first birthday, and he can see to read without spectacles ; still rises early, was up this morning at four ; fancied he might be the father of the Society.

1864.

September 24. To Earl's Colne . . . Mr. Cawardine told me he remembered having seen Dr. Johnson. He was walking at four years of age through St. Paul's Churchyard, holding by his father's finger, when his father, pointing to an old man dressed in a snuff-coloured suit and worsted stockings, who stood as if resting by a post, said, "That is the great Dr. Johnson." Mr. Cawardine also remembers Cowper.

November 30. Our Anniversary Meeting . . . not over till 6.15. Seventy-eight sat down to dinner at Willis's . . . Sir Charles Lyell responded to the toast of. Darwin's health; he was scarcely heard at our end of the table, but he began by saying that twenty-seven years ago when the *Beagle* was about to sail, some one recommended Darwin to read his (Lyell's) book, but not to believe a word of it. This some one, Sir C. told me afterwards, was Henslow, and that in later years it had been said of him that if he lived to be sixty to see his books in a seventh or eighth edition, he too (Sir C. L.) would be an obstructor of science.

Tyndall's speech was good, head and heart, but the way in which his health had been drunk, had deprived him of head and left only heart; that the R. S. had been supplying him with everything but brain; had given him money from the Government grant, had made two of his papers Bakerian Lectures, and had now crowned their work with the Romford Medal. Eleven years ago he was associated with Darwin for one and a half days, now associated again, but with a medal, and on higher footing, concluding with eloquent eulogium on Darwin.

December 2. Dr. Gray called for a talk, approves of what the President said about the 'Origin of Species,' wishes the R. S. had no medals to give away. Mentioned the quarrel raised when a Royal Medal was given to Mr. Beck; said that was a job; that Dr. Todd and Mr. Bowman were determined Dr. Lee should not have a medal; that the Council referred the whole question back to the Committee; that Dr. Roget, Secretary, laid before the Committee a false minute to the effect that they were to give their reasons for their award; that he (Dr. Gray) wrote a full account, with much evidence, to Lord Northampton, and begged him to lay it before the Society; that his lordship said he would do so, but that at the same time he must give in his resignation, on which he (Dr. G.) exonerated him. That he shall propose to print the great catalogue in sections according to language. That when he gave his evidence, and printed his pamphlets about the catalogue and management of the British Museum, of which he gave me a copy, the late Earl of Ellesmere sent Dr. Cureton to him to say that if he (Dr. Gray) would withdraw all his evidence and charges, he, the Earl, would invite him to dinner.

1865.

May 14. Yesterday returning from my afternoon walk, I met Dr. Livingstone; he had on his gold-banded consular cap. Told me he was going to Scotland to prepare for another journey to Africa. He introduced me to Captain Grant (of Speke and Grant fame), who was with him. A tall, quiet-looking, gentlemanly man. He is about to return to India.

May 19. With Professor Stokes to the banquet, given by Lord Mayor Hale, to President and Council of the Royal Society; nearly 300 present. As a whole, the dinner was not good, not equal to Royal Society Anniversary dinner at Willis's.

July 4. Had an evening party for music and dancing, forty present. Among the ladies were some very pretty faces; the Miss Mulocks, relatives of the authoress, were much admired.

November 8. General Sabine called preparatory to his going to St. Leonard's for a month. He talked of his forthcoming anniversary address—

o

of the topics to be noticed. I suggested Mr.
De la Rue's lunar photographs, and the utility
of photography to astronomy. He adopted it.
He spoke of the Preface of the Great Catalogue
—whether it would be published in one volume,
in 1866, and who should draw it up, and who
sign it: whether himself. Then of the papers
he has in hand—of the Presidency, and who shall
be his successor. He thinks Professor Stokes
the fittest by reason of scientific ability, but
says he has not governing faculty—besides his
pecuniary resources are insufficient. Mr. Graham,
Master of the Mint, would do, but his health
is too feeble. At the time General Sabine was
talked of for the Presidency, Mr. Graham, who
was also talked of, wrote to him saying, "If I
accept it, I shall not live six months." "Would
Sir Philip Egerton take it?" I asked. "No,
never. He never opens his mouth in the
House. He has a great dread of public speaking.
Refused point-blank to be President of British
Association on that account—refused even to
second a motion. His conversation besides is
controversial, and manner abrupt. Sir R. Murchison is not sufficiently discreet and judicious.
He (the General) thinks there must be a crash
at the Geographical ere long."

December 5. Took my first ride out with Riding-Master Lumley, round the Regent's Park. Found less difficulty and more enjoyment than I expected.

December 7. Yesterday at Woolner's. I saw his bust of Gladstone and the plaster model of Carlyle—both excellent. Both thoughtful, yet how different in the expression.

In our after-dinner talk Woolner spoke of Carlyle, who on the question of opening museums on Sunday said, "He would be sorry to give the old religion its last kick." The party much amused by my saying it had been remarked that the old religion, meaning the Church of England, might be cited as an instance of a set of harness continuing to go long after all the animals that once were in it had died.

1866.

January 7. Mr. Huxley gave a lecture at
St. Martin's Hall in the evening to a crowd,
2000 people present, as many more went away.
Even the half-a-crown seats were all filled.   The
text was the Great Plague, the Great Fire, and
the Royal Society.   A bearing of one on the
other was shown, and the Lecturer stated that
if all other scientific records in the world were
destroyed, the Philosophical Transactions would
be an excellent nucleus with which to make a
new start.   I did not hear the lecture; these
particulars were given me next day by Mr.
Huxley and Mr. Busk.

January 26. While the Electric Diving Light
Committee was sitting, Sir C. Lyell called to see
Mr. Huxley.   He talked about the St. Martin's
Hall Lectures; that as a pecuniary speculation
they pay, and the promoters will probably have
£150 in hand on completion of the Six.   That
Dr. Carpenter was in his lecture very aggressive,
animadverting on Scripture, denouncing some of

the Psalms as fiercely immoral, that man lived long before Adam, that nine-tenths of the clergy taught what they did not believe. He, Sir Charles, sits on the platform, says many of his friends have written to warn him against taking part in such proceedings; that for his part he thinks it would be better to be less aggressive; that he fears James Heywood will make a mistake scientifically as well as aggressively.

We spoke of Carlyle. Sir Charles said he once met him at a party, where Carlyle at the far end of the table discoursed so long on silence that no one else could get a chance to be heard. That he (Sir C.) on getting near to him afterwards, told him so, and that Carlyle received the hint with good-humour.

March 14. R. Merington continues to give me a weekly lesson on the piano. Three months of practice have somewhat diminished my awkwardness, but it will be long before I shall be able to count time and play with facility.

March 15. At our evening meeting Mr. Jno. Evans' paper on a possible geological cause of a change in the axis of the earth's crust was read. It attempts to account for changes of climate; a swelling in any part would work down to the equator, a cavity would rise to the pole, as was shown by a model. The theory seemed acceptable, but the facts were questioned. Mr. Grove argued that the slipping of any part involved a slip of an entire zone of the crust. Prof. Ramsay said the theory would account for some geological phenomena, but that the great periods of glacialisation were independent of it. Mr. Mallet denied the glacial periods, and disputed the proof as shown by grooved rocks, and showed that no change of axis was needed to account for change of climate: the poles cooled first down to a tropical temperature, then life appeared, then this temperature progressed gradually down to the equator, leaving its traces in the fossils of all the zones through which it passed. Mr. Spottiswoode hoped to have before long a mathematical demonstration of Mr. Evans' theory.

March 16. Sir Charles Lyell called to look at Mr. Evans' model, and to enquire concerning the discussion. Spoke of Professor Stokes' objection, that plants would not sleep through the long Arctic winter, to which I said that if there had been a shift of axis, it (the axis) may have formerly been perpendicular, which giving the same equinox to all parts, would get over the objection to the long polar winter.

He spoke of our movement through space, and said he had heard recently as a proof of our approach to Hercules, that the stars in that constellation are beginning to open.

April 15. Walk to Sheen Lodge, a cheerful welcome from Prof. and Mrs. Owen, a Mr. Kilgour, and Mr. and Mrs. Henry Cole. . . . At dinner Cole gave us an account of Whymper's siege of the Matterhorn, carried on for years, circumvented by the Italian guides, but accomplished at last.

July 23. To Hyde Park after 4 p.m. Near each gate was drawn up a company of police and a small detachment on horseback; the largest force was at the Piccadilly entrance, all in preparation for the Reform meeting announced to be held in the Park, but forbidden by Mr. Walpole, Secretary of State. A good many roughs were in the Park, and the Achilles statue and slope beneath were occupied by ragamuffins; respectably dressed people however were most numerous as usual. At 5 p.m. the gates were closed, any one might go out, but no one was allowed to go in. At 5.45 when I walked home, there was a great crowd round the gate, and a stream of people was setting down Piccadilly towards the Park. No disturbance then; a man in the Park was selling his lollipops under much less excitement than at a review.

November 11. To Dulwich to dine at Cresy's. Mr. and Mrs. Hill of the Dulwich School were there, and Miss Garrett, the doctoress : the latter a very agreeable person, fair, bright hair, very intelligent expression set off by a black velvet dress. At times her style of speech is somewhat short. I accompanied her back to London ; on the way she told me that since July she had had 1400 patients, women and children, at her dispensary. . . . She likes mathematics and astronomy, and if an opening had offered might have followed one or other as a profession.

November 13. To Clapham Junction. Walked thence to the Common to a meeting of the Debating Club at Bucklands—opened with the proposition, "Does 'Ecce Homo' contain a fair exposition of Christian thought ? " Praised the book, bright, enthusiastic, breezy. A clergyman found faults. Chairman praised. Mr. Newmarch, in excellent speech, showed that the book was opportune—fell like spark on tinder—refers matters to man's own heart and conscience, is therefore opposed by those of attenuated forms and mere religious ceremonies. Nothing new in it : has all been said in Tucker's ' Light of Nature,'

Cudworth's ' Intellectual System,' and Hartley on Man. I held my peace, being unwilling to disturb the harmony of the evening.

November 27. In the evening a dinner party. Miss Garrett, *i. e.* Miss Dr. Garrett, Miss Delf, Mr. and Mrs. Cresy, Mr. and M. Merington, and S. Hamilton. Much pleasant talk ; most of the ladies of opinion that there is no difference in the male and female intellect if trained alike. Some talk there was as to whether Christianity or gunpowder has done the more mischief in the world.

November 30. Our Anniversary, satisfactory meeting. Not so many as usual at the dinner. Sir Stafford Northcote spoke, and showed that the Government lay under great obligations to the Royal Society for advice on scientific subjects.

December 1. The President called and talked for two hours. Pleased with yesterday's result, with the prospect of the relation in which the Society is to stand towards the State.

December 9. Drove to the British Museum to dine at Dr. Gray's. Various talk about the Museum, Sir F. Madden's resignation, and Dr. Gray's purchase of a Natural History collection in Paris. The Blacas collection just bought was examined by Newton, who telegraphed to Disraeli, "Worth £43,000." "Buy it," was the answer by telegraph. Next day Newton arrived with a number of the gems in his portmanteaux. A French Commissioner came to examine the collection, when he had finished viewing it, to report thereon to the Imperial Government.

1867.

January 5. A talk with Dr. Sharpey about
the new Meteorological Committee, the mori-
bundity of the British Association, and the new
home of the Royal Society, which, as Banks and
Barry now state, is to be built on three sides of
our court-yard, with a façade to Piccadilly.

February 8. To sister Jane's for the evening.
Two Methodist preachers there, Dr. Waddy and
Mr. Maunder, and their wives, and half-a-dozen
elderly ladies who scarcely talked. The men,
including Mr. Corderoy, spoke of the Power of
the Keys, not believing in apostolic remission of
sin; about Reform and the growth of democracy;
how to put down the roughs; about preaching,
preachers, and prayer-meetings; the opinion being
that the two most useful preachers at present in
London are Spurgeon and a Baptist butcher
who has built a chapel at Shepherd's Bush.

March 29. To-day the last of the trees, the
old elms on our lawn [at Burlington House], was
felled. For five years I have seen them bud and

leaf in spring, have heard the starlings, thrushes and sparrows; now all is havoc. The University building has risen above its basement. Nearly all the trees were hollow.

April 16. While potting ferns before breakfast was accosted by Mr. Eyre, registrar of the Academy, who comes to see how the excavation of foundations goes on. He will one day occupy my rooms, and does not like the prospect of their darkness.

May 16. Spoke to my chiefs about going to Norway and taking July instead of August. They approve, should no business come in to prevent it.

May 18. Called on Miss Plessner about lessons in Norwegian.

June 22. The President called this afternoon on his way home from British Museum. Said he was in trouble about the Copley Medal, for which Regnault has been nominated three times, and now Mr. Grove suddenly proposes Mr. Wheatstone. The President complains that he cannot get the officers to act as a Cabinet, they prefer to act as individual members of Council. For himself he would prefer that Regnault should get the medal this year, and Mr. Wheatstone next year.

Mr. Spencer Walpole called to pay his subscription, looking stout and well. I remarked,
"He was no doubt enjoying his holiday out of
the Home Office." "That I am," he answered.
Then I—" I wonder they get anybody to take it."
"You may well say that. It.is a great worry.
No one knows how much but those who have
held it." "Besides all the titillation by the
newspapers." "Ah! they always argue wrong,
they don't know the facts."

September 26. Dr. Sharpey came in just
returned from his holiday. While he was with
me Dr. Hirst and Mr. Khanikoff came to com
pare some of the Newton-Pascal documents
with our collection of Newton's letters. The
documents were scraps and half sheets all
written in French and signed I. Newton. On
strict comparison it was easy to see the writing
was not Newton's. Mr. Khanikoff stated that
Chasles supposed the mass of documents to be
part of a collection made by Des Maizeaux, while
resident in London. Thomson's book gave the
date of Des Maizeaux's election as F.R.S. I
fetched the Journal Book, found the entry of his
election and admission, and a prior one of a
present made by him to the Society, October
1720, his 'Recueil de pièces diverses,' in the

second volume of which I found the original of four out of the six documents before us; three being paragraphs from the French translations of Newton's letters, and one a paragraph from a letter of Dr. Clarke's, though in the document this was signed "I. Newton" like the others. This was conclusive as to the forgery. Khanikoff goes back to Paris thoroughly convinced, and Dr. Hirst is to write a statement to 'The Times.'

October 4. This day Miss Hills recommenced my music lessons.

October 19. Our porter (at 2 p.m.) brought me up a pasteboard luggage-label, bearing the name of Henry Morris of Philadelphia, who wished to see me. Presently entered a man . . . with huge projecting ears, who told me he had been to Chapman and Hall to enquire for my address, as he had read all my books, and had a strong desire (if the author were not a myth) to shake hands with him. He first saw the books in the Philadelphia Library, then ordered the whole series for himself and read them eagerly, as did also his family. He hoped I should yet publish others. He had a large iron foundry, with 500 hands; when his sons became twenty-one they said, "Now, father, you go out of this and leave us to run it," and now they are running it and employ 1100 hands, and when he walks through he sees how much better managed it is than in his time. So he is not one of those who thinks the old ways always the best. Now being free, he has come for a travel in Europe. He stayed about three-quarters of an hour; we had a good talk, and then with a hearty handshake we parted.

October 23. Banks and Barry called to talk of the difficulty they have with the Albany as to the obscuring of lights by our new building, and to ask whether the Royal Society would take west wing instead of east. If so, they could begin to demolish and pull down at once.

November 22. With help of T. Wheeler, finished balance of my account-books this evening, two evenings' work; shall now be well prepared for the auditors. Then played a peal on the piano and read.

December 3. To Buckhurst Hill; gave a lecture on Norway to the school. No cab to be had at Shoreditch on return at 10.45, owing to the strike of cabmen in consequence of the regulation requiring a lamp.

1868.

January 11. Close work in office, as, in addition to the routine, I am writing biographical notices of deceased Fellows for Dr. Sharpey. The Doctor was well enough to come to our meeting last Thursday, and was heartily congratulated on his recovery. Mr. Babbage came, a rare incident.

February 1. Our President talked about the successor to be appointed to Sir Richard Mayne ; spoke of Mr. Knox, police magistrate, as a likely person. Knox used to write for ' The Times,' at 8.30 p.m ; Mr. Delane used to give him and others their task and order for leaders. At times, owing to change of circumstances, this had to be re-written, and at times a third leader on another subject was required. The exhausting effect occasioned by having to write the second and third under pressure Mr. Knox describes as extreme.

June 3. Clerk of Works brought Gustave Doré to look through our rooms. The famous artist is compact, about 5ft. 6in. in height, and like the photographs one has seen of him. He speaks no English, appeared to take interest in the things I showed him, and said he would perhaps come to an evening meeting.

June 8. Sir H. Holland called to appoint a meeting of the Faraday Monument Committee. Told Dr. Sharpey he saw Disraeli yesterday, that he (D.) is favourable, that they spoke of Hallam's statue, that Gladstone and one or two others had proposed a Latin inscription, that he (Sir Henry) had maintained the contrary and carried a vote against Gladstone. " I wish to God," said Dizzy, " I could carry a vote against him."

October 12. The Colonnade in our fore-court so much talked and written about is to be stacked in Battersea Park, and this day the taking to pieces was begun.

October 24. In talk this afternoon our President said that one principal reason why the Royal Society stood better than ever in public opinion, was the diligent way in which the business of the office was carried on : that this result could never have been achieved in my predecessor's time.

October 31. Another talk with General Sabine. He imparted his views as to the way in which the new Council list should be prepared, viz. by

P

President and officers, then the Council might suggest alterations. That he would not be President to have to fight a faction, that he talked to me of these things, because he would not be much longer President. That the Duke of Argyll had been asked if he would be President, and had declined.

November 26. Our second meeting in our new temporary meeting-room; complaints of its being too warm. A most interesting evening; Lockyer's paper, 'Spectroscopic observations of the Sun,' was read. It occasioned a discussion in which Mr. Babbage, Astronomer Royal, Mr. De la Rue, etc., took part. They all considered that Mr. Lockyer had made out his case, and praised him accordingly.

November 30. Our Anniversary. The room quite large enough and to spare. The President well heard. His speech a capital summary of the year's science. Sir C. Lyell moved, and Sir T. Watson seconded the vote of thanks. At the dinner were 70 guests, including Lord Justice Wood, Dean of Chichester, Sir Bartle Frere. Sir C. Wheatstone returned thanks for the Copley, Wallace for the Royal, Stewart for the Romford. In proposing the President's health, Lord Justice Wood said that of the three maces, the Commons', the Chancellor's, and the Royal Society's, the last was the one which represented proceedings of most benefit to mankind.

December 16. To Buckhurst Hill. Conversational lecture to the young ladies on composition, and criticism of their narratives of a journey.

1869.

January 15. Dr. Tyndall's opening lecture at Royal Institution, 'Effects of light on vapour, formation of cloud.' A laugh when he spoke of having been to the summit of Primrose Hill; and that the audience disturbed the "purity of his thoughts" (serenity he meant).

February 2. Had a large party of twenty-two. . . . Dancing went merrily till one in the morning.

February 13. With Marcus Stone by invitation, to Clunn's Hotel, Covent Garden, to a meeting of "Our Club," which meets to eat, drink, and talk nonsense. Hepworth Dixon, Dr. Percy, C. Dickens, jun., and about a dozen others were present. At eleven o'clock migrated to the Arts Club and looked on while Marcus and his friend lost a game of billiards.

April 6. What a scene to-day with the bringing of pictures into the Academy! Those by hand and in cabs were delivered at our front-door; the corridor, and soon half its length on each side, had pictures leaning against the wall. The back entrance was for vans, and there from 6 p.m. to 2.30 on Wednesday morning the scene was sensational. Vans came in, the pictures loaded into common trucks, all too rudely handled as it seemed to me, and were drawn to the lift and sent slowly up. Then the vans drove across the temporary plank road to Piccadilly. The work was slow and without method; artists grew impatient, brought in their pictures, placed them against the wall, and soon there was no room for the trucks on the floor of the great passage. That much damage was not done is marvellous. And among the "artists" how many looked shabby and dejected! . . How is it so many men mistake their vocation?

April 10. Went to look at the limbo of rejected pictures. What a sight! looks like 1000, and the water-colours not sent down yet.

May 1. Another before-breakfast walk through the Academy. One of the porters told me that last Wednesday the Hanging Committee and Council did nothing but eat and drink from 1 p.m. to the close of the day, which the head porter verified.

May 12. The Queen came at noon to see the Royal Academy, Princesses and Princes with her. Sir F Grant walked by her side, his height made her look the shorter. Except in a carriage I had never seen her before, so she appeared to me very short. . . She stayed an hour.

June 4. To tea with Dr. Bird . . . later Swinburne came in, a small, weak-looking man, jerky speech, fidgets and fluttering of hands when excited with talk. Told us he was writing a review of Hugo's 'Homme qui rit,' for the 'Fortnightly,' and recited his last sonnet on the "Accursed," as he calls Louis Napoleon. In one of his lines he names him "Iscariot the Second."

June 24. To the Royal Academy *soirée*. The galleries looked remarkably brilliant under the gaslights, and the pictures could be seen to advantage, but there were too many scraggy necks and shoulders and too many bony arms.

June 29. To an "At Home" at MacCullum's, a picturesque gathering, but too quiet. . . The painting was hung in the back room, for which the 'Spectator' called MacCullum "The Wordsworth of Art."

July 31. Started for Harwich and holiday. Luxembourg, Treves, Vienna, Dolomites, etc. etc., and back to London.

September 15. Stormy passage across the Channel. Soon after my return J. N. Lockyer

called to propose to me an engagement to write
for Macmillan's new weekly periodical, ' Nature.'
I decline a regular engagement, but agree to
write occasionally.

October 7. To Windsor with Cresy. Met Mr.
Woodward at the Castle by appointment. He
showed us rare works of àrt in the Queen's
Library, old Italian Masters, old illuminations,
part of the loot from Magdala : curious to see the
Ethiopian eye and style of feature. . . We were
merry, none of us anticipating that poor Woodward would be dead on the 13th.

1870.

April 19. Mr. De la Rue says Mr. Grove is
known as "Shady Grove" because of his
hypochondriasis. Our President shows signs of
the severe though short illness he has passed
through.

May 31. To Mr. Vignole's *soirée* at Civil
Engineers' . . . talk with sundry acquaintances,
G. T. Reed, Hepworth Dixon. Latter says
' Pall Mall Gazette ' will soon die, that Smith has
lost £41,000 on it. He (Dixon) intends to bring
out a paper, and promises to have a talk with me
thereupon by and by.

June 9. Was able to-day to gratify a long-felt
desire, and went to hear the Charity children sing
in St. Paul's . . . no crowd, could see the whole
range, the slopes of children under the dome,
white and black. The effect of the 5000 voices
singing the Old Hundredth was grandly
impressive. It thrilled through me and brought
tears into my eyes.

June 26. To the Toynbees'; pleasant house to

visit. Arnold talks well, and appears to be a promising youth.

June 30. Sir C. Wheatstone tells me he bought some books at the sale of Sir J. South's effects; among them was Dr. Granville's ' Science without a Head,' on which Sir James had written "By a Head without Science."

July 31. Rail to Dover, steamer to Calais, cloudy and calm. The French captain talking with his compatriots admits possibility of a "premier insuccès" in the war with Prussia, but in that case it will be all the worse for Prussia afterwards, "une guerre d'enragés."

October 1. Sixteen years ago the book [of diaries] preceding this was opened. The years since elapsed take a large piece out of my life, but experience lengthens as years shorten, and the pleasures that attend on grey hairs take the place of the active and boisterous pleasures of earlier years. Nature is kind, and so we accept the change without regret or even with satisfaction. Work is still a pleasure to me. I have never yet wished that I had not to do my daily task. Travel and natural scenery still afford me delight, and at need I can walk twenty miles. Much material have I accumulated that could be worked into an interesting book, but I

doubt if it will be written. Subjects that occur
to me are Hull to Hammerfest; chapters of
travel; among the mountains. Besides which I
should like to print (not publish) my poems.
My brothers and sisters and some of my friends
would like to read them in a book. At present
the ballad ' Prisoner and his Dream ' occupies my
attention. My relations with my chiefs and with
the F.R.S. generally continue as agreeable and
friendly as they always have been.

October 10. Another gap made in the number
of our Council by the death of Dr. Matthiessen.
He killed himself with prussic acid in his
laboratory at St. Bartholomew's Hospital. He
was a highly sensitive man, overworked himself,
had no firm moral courage, and was worried
beyond measure by his class, hence in a moment
of trial he cut short his life rather than face
difficulties. He had told one of his friends that
he could do nothing with his class except by
treating them as dogs. Dr. Duncan tells me the
doorkeeper of his class-room at King's College
said to him, " You mustn't think, Sir, that 'tis the
same as when you were a student here. 'Tis
a different class. 'Tisn't for the knowledge that
they come now, but what they can make out of
it. In your day, Sir, the young fellows were

gentlemen, now they are not." Everything confirms my opinion that what is wanted is not more appliances for education, but students who are really in earnest. The President says that Mr. Vaux of the Museum has resigned, another case of mental derangement.

October 15. Grieved and pained to read in ' The Times' note of the death of my friend Cresy. He was a late acquaintance, but I liked him heartily and shall miss him painfully. On September 29 he was at his office for the last time, complained of pains which proved to be the precursors of an attack of small-pox. The seeds of this he had doubtless picked up in his journey through France : he was five days coming home from Geneva. How *memento mori* grows upon us with our years !

October 23. Again rain puts off my visit to Professor Owen at Richmond. . . The Middlesex magistrates have refused dancing licenses to the Alhambra and Highbury Barn. It was time. Are we to adopt and imitate all the depravities invented in Paris, and so deserve more and more the judgment that has fallen on France ?

October 27. First Council preparatory to our Session. Excitement as regards choosing new Treasurer and new blood, including younger men into the Council. Mr. Gassiot and Mr. Spottiswoode were proposed as Treasurer; the ballot was even: the decision will be made next Thursday. President spoke of his intention to resign in 1871, he will then have been at the head of the Society ten years. He greatly wishes to have Mr. Gassiot as Treasurer for one year. The sitting was late, and with helping Dr. Sharpey to finish the minutes I did not get downstairs to my dinner till 7.45.

Professor Miller of Cambridge says that students of the present day behave better than those of thirty years ago, but have not the same strong desire to learn. He agrees with me that love of learning is more likely to secure material benefit than study taken up merely for a living. He mentioned the late Sir B. Brodie, who used to say he would rather see medical students with a mere knowledge of details of their profession and an acquaintance with other branches of knowledge, than a profound knowledge of anatomy with nothing else.

November 1. Met Professor Ramsay in the Park. He says it is not true that there is a desire for knowledge and for study, except among working men, these crowd his lectures.

News that Sir G. Sabine will next year resign

the P.R.S., and that Mr. Grove is to be his successor.

The President says to me in a note to-day: "I look forward with regret to our ceasing to work together for the advantage of the Royal Society," which implied that he meant to resign. However, that conclusion is not yet come to. He did not come to the Council, and left them without restraint to nominate Mr. Spottiswoode for the vacant place of Treasurer.

November 22. Dr. Wallick told me to-day that Sir Roderick (Murchison) was struck with paralysis yesterday and lies dangerously ill. It was but Thursday last the grandiose baronet gave notice of proposing the Duke of Sutherland for election. Should he die the Geographical Society will lose much of their prestige. Wallick was once speaking to Sir R. of what folks said of his photograph, it looked too tame. "Ah," answered Sir Roderick, "you should take me after dinner when I have a bottle of port in me; I look sprightly enough then."

The President sent me a letter of Mr. Gladstone's, from which to notify the Council that the Prime Minister will come to our Anniversary Dinner should he be in town.

November 22. Mr. Gassiot called to show me a

proof of a pamphlet (which he intends to publish) concerning the controversy about the nomination of Treasurer: the imputation that he (Mr. G.) coveted the post; and the blame thrown upon the President. I answered that so far as he himself was concerned there was no need of publication, because all who knew him would acquit him of a desire for the place, and that as regards the President, the majority knew full well that the Society, in all their existence, had never had a better, or one that had done so much to lift up their reputation, and make them the head of science in Europe. Moreover that such strife should be left to die out for want of fuel, and that "Woe unto you when all men speak well of you," was a precept worth keeping in mind always. He replied that he should publish notwithstanding, and explained how it happened that he had laid himself open to the charge of coveting the post of Treasurer. Some three years ago Dr. Miller wished to give up the post and asked him to take it. "I refused," he continued, "and always among the Fellows my acquaintances, made known that at my years it would be too heavy a place; and having felt warnings that I must take care of myself, I had promised my family that I would not accept it if offered to me. But last month I received a letter from the President urging me as a matter of personal friendship to take the place for one year, which would be his last year of office. I felt it would be difficult to refuse, and showed the letter to my wife, and she said, ' There is no help for it, you can't decline.' That's how it was ; but I felt it would be a severe year for me : not with the finance, because that gives me no trouble, but with the general administration of the Society's affairs."

Mr. Gassiot looked ill, and I noticed that he dropped here and there a syllable as if from partial paralysis. He told me he was under strict orders from his doctor to keep quiet, and to refrain from reading and work of every kind.

November 24. At our evening meeting Dr. Carpenter gave an exposition of results of his cruise in the Mediterranean, last summer, in the ship *Porcupine.* It was clear and logical, though too long from his habit of repetition, and fully made out the difference of density, or specific gravity, of the Mediterranean water compared with that of the Atlantic, and the existence of the deep-down outflowing current.

November 30. At our Anniversary Meeting to-day the new Council was elected unanimously,

and Mr. Spottiswoode is now our Treasurer. He
will, I fancy, prove acceptable to every one. At
the close of his address the President announced
his intention to resign in a few graceful and
touching words. At the dinner (at Willis's) he
had Mr. Gladstone on his right, and his old
friend the Lord Chancellor on his left. My
desire to hear Mr. Gladstone speak was at length
gratified. He made a good speech : had hoped
that for one evening he might have been allowed
to forget his own consciousness, and the weary
weight which Ministerial responsibility in-
volves. But it was not to be, and he spoke of
the relations between the Government and
science, as represented by the Royal Society, and
said that Government were always glad to
entertain any reasonable proposition in behalf of
science : that the recent misunderstandings as to
the Eclipse expedition had arisen from the
astronomers having been so much taken up with
celestial matters, that they had failed in some of
their terrestrial proceedings. Two centuries ago
a student could master the whole sum of human
knowledge and would naturally acquire great
breadth of thought ; but now in science as in
manufactures, division of labour prevailed and
narrowed the scope of thought. This result,

though to be deplored, was nevertheless inevitable, and must be accepted, but all, whether handworkers or headworkers, should strive to associate breadth of view with their employment. The Lord Chancellor gave a sketch of the President's services rendered to science, and said he hoped the days had gone by when the Society would be content with a mere nobleman as President.

December 15. Mr. Gassiot came to the Council; he still looks ill, and has to avoid excitement. He told me that he had received many letters from Fellows approving his pamphlet; among them one from Lord Salisbury. I told him of four, Sir J. Alderson, General Boileau, Dr. Hoskins, and Mr. Mallett, who expressed their approval warmly.

December 22. At our evening meeting, after the reading of Archdeacon Pratt's paper on the 'Constitution of the Solid Crust of the Earth,' Mr. Siemens questioned some of his views, and described an instrument by which he could determine the gravity of the "crust" below the sea. The instrument is floated and the result read off on a scale at the upper end. In the Bay of Biscay, at a depth of 2500 fathoms, the error would not be more than two fathoms.

December 30. ——called, also Dr. R. Copeland
from Göttingen, who was astronomer on board
the *Germania* in the recent North German
Arctic Expedition, and is now on his way to
be assistant to Lord Rosse. He showed me a
sketch map of the coast of East Greenland, laid
down up to 77°, and the outline of the Shannon
Islands rectified, and the course of the deep
fiord into which the expedition penetrated. He
recognises the merits of Clavering's survey, great
part of which needs no rectification.

He gave some particulars of the quarrel between Petermann and the organising Committee,
and officers of the expedition: it originated in
his changing the name Dove fiord to Franz Josef
fiord, and in the questioning of his maps by the
explorers.

Concerning the war, he says the Germans are
getting tired of it. That another army of
200,000 men is to be sent into France in the
beginning of 1871.

The writer of the saucy reply to the letter of
the Royal Irish Academy, in deprecation of the
bombardment of Paris, was the Pro-Rector Dove,
son of the Dove of Berlin. Copeland says if
the father had written it, it would have been no
less severe but more courteous.

There are commonly 800 students in Göttingen, one half of them are now at the war. The number used to be thousands, but the Government having in 1837 required the Professors to take a rigid oath of loyalty, the best of them have emigrated, and since then the numbers have never reached 1000.

December 31. To the private view of the Academy Exhibition of Old Masters. Landscapes by Crome, Constable, Hobbema, eight or ten Murillos, and the boy portrait by Gainsborough.

Met Dr. Percy, who told me an attempt is to be made to deprive the paid officers of the Royal Society, who are Fellows, of the right to vote.

1871.

January 4. The President called, talked of yesterday's interview with Dr. Copeland, and mentioned that the Germans think of undertaking this year an Arctic dredging expedition with deep-sea soundings, and tracing of currents. He thinks they should endeavour to measure an arc of the meridian from a five-mile basis in Greenland.

Next he showed me a copy of the letter he had received from the Science Education Commission, indicating the information they would like to receive from him when they call upon him for evidence, and wishes me to furnish him with the requisite documents.

Then referring to his having announced his intention to resign the Presidency, and the proceedings in Council preliminary thereto, he said he had received many very friendly letters thereon. The Marquis of Salisbury expressed himself strongly on the rude scene in the Council of November 3, and Mr. Mallet wrote that the phrase " infusing new blood " meant infusing those who used the phrase.

The President will take the chair at evening meetings, but not at the Council meetings during the present session (excepting the Council for the medals). He thinks he was not well treated by the rejection of his suggestion that Prof. Owen and Sir C. Wheatstone should be chosen into the Council : his reason being that in the selection of a new President, the opinions of two men of so much experience would have been very valuable.

He looks forward with pleasure to the opportunity that his retirement will bring, for the finishing of his labours on Terrestrial Magnetism.

His scheme for the succession was to ask Lord Salisbury to take the Presidency for two years and then transfer it to Lord Rosse.

January 9. The President called. He spoke of Mr. Grove as one who might make a pretty good President : but that he hears he has become too fond of sneer and snarl.

January 12. Mr. Marshall brought the American Leland, author of the ' Hans Breitmann Ballads,' to our evening meeting. He is a tallish man, with sandy hair and beard. In profile his face appears somewhat Coptic. His expression is manly and quiet. After the meeting I showed him our Charter Book, which he said was far more worth looking at than the Golden Book at Venice. The signature of Sir Kenelm Digby attracted him : he said he had read the old Knight's works.

January 19. Sir James Alderson, at the R.S. meeting, told me he should like to get up a round robin to the President to beg him to retain office another year, which would give him time to inaugurate the new House.

Sir William Thomson explained a new method of estimating resistance by one deflection of a galvanometer, and a constant form of Daniell's battery, and a method of determinating a ship's place at sea without the usual calculations and logarithms.

January 21. Mr. Gassiot called to speak of his project of reviving the Scientific Committees. His chief reason is that they would save the new President much work ; that if Presidents are to be elected for two years, some of them will be men who will like the honour but won't like the work.

January 26. Prof. Williamson at our evening meeting gave a singularly interesting exposition of the Calamites of the Coal Measures. In the tea-room talk afterwards, I heard that there is an intention to set aside others of the chief officers as well as the President.

February 10. Mr. Clare called, and referring to Mr. E. J. Reed's paper read last night, said that Reed had got all his ideas from him; that he (Clare) was the inventor of the *Warrior* and other improved ships : that the Committee now sitting would show Reed up.

February 22. Dr. Stenhouse called. He is very sore against the Chancellor of the Exchequer for abolishing so summarily his place as Assayer to the Mint, thereby depriving him of £600 a year. The fee for an assay is 2*s*. 6*d*. : the sum-total of fees in January last was £147.

The Doctor said Binney, F.R.S., had called on him and told him a story. He, Binney, and two others, were going to the meeting of the British Association at Cork. They placed their coats and rugs on three places on fore-seats of the coach, and went into the inn to breakfast. When they came out they found two of the places occupied. They remonstrated : the intruders refused to budge, "they were going to the meeting of the British Association." "So are we," answered Binney, and being a big powerful man he threatened violence. The two delinquents

(Tyndall and W. Francis, printer) then gave up their seats.

March 11. The President's *soirée* went off satisfactorily. Not so many articles as usual for exhibition, which was an advantage.

March 13. To dine at Bird's; Ionides and his wife present, and Mrs. Lynn Linton, a bright and clever-looking woman. We had a talk, and agreed to have another, and pursue the topic as to free-will and morality. Met also Edwards the artist, who talked a good deal about Leland, author of 'Hans Breitmann Ballads.'

March 15. Mr. Siemens called. He is of opinion that our evening meetings would be better attended than at present, and more lively, if the discussions were printed.

March 17. Our Treasurer and Prof. Stokes went to Greenwich to offer the P.R.S. to Mr. Airy. He accepted without reserve. The offer was supported by an unanimous vote of the Council.

March 18. News in 'Times' of Robert Chambers' death; set me thinking of 1844 and his kind advice and sympathy when I began to write for the 'Journal.'

March 30. Leland, the 'Hans Breitmann,' was at the meeting; I had a talk with him. He

likes England greatly, finding so much of culti-
vated society. I put to him a query I have often
put to Americans, and for the first time got an
answer, viz. What is the reason of the passion
for Paris that prevails among Americans ? Rich
New Yorkers, he answered, not caring for intellectual culture, but for show, and gaiety, and
sensational life, find what they want in Paris with
great facility for spending money. For ordinary
times a colony of about ten thousand is settled
about the Arc de Triomphe. They can get into
the highest society, and their daughters can
marry titles if they please. In England, on the
contrary, they are always overshadowed by a
class into which they cannot penetrate, are
oppressed by the weight of some one above
them. Mrs. Sandford, wife of the American
Minister at Brussels, says she can give her
parties there and be equal to the highest, while
here she had always a feeling of being kept
down.

April 4. Met Dr. Percy at our club. He is
indignant at the report of the Science Education
Commission in which they propose to remove
School of Mines to South Kensington. Says
the whole enquiry is a job, got up by Dr. ——,
who wants an endowment for University College.

That Huxley is playing fast and loose, and is planning to get a good post for himself.

Our President tells me that Adams is revising Gauss' 'Theory of Terrestrial Magnetism,' and that Kew will be a good place for the computers who will be required.

April 15. Dr. Carpenter says he has known Dr. Percy thirty-five years, and never yet heard him say a good-natured thing of any one.

Mr. R. Mallet confirms the rumour that the Science Education Committee is a job, got up with a view to increase the emoluments of —— at University College. Dr. Percy's article in 'The Times' and his letter reflecting on Mr. Huxley are condemned.

The President tells me that Mr. Airy, in view of becoming P.R.S., has written to the officers of the R.S. to ask that £100 a year may be allotted to him for travelling expenses, or expenses incurred in Presidential duty. The President says further, that the Government had in contemplation the giving of life peerages to some six or eight scientific men, with a view to liberalise the House of Lords, and that in all probability the President of the Royal Society would be one of the number.

April 16. To dine at R. Mallet's, Clapham

Road ; Dr. Haughton of Dublin there and two sons. Much talk about Ireland. Haughton says the Irish regard Gladstone as an orange, and when they have squeezed Papist Denominational Education out of him, they will throw him aside. That in five years from the time the priests have their own way with education, an Irish parliament will be sitting in College Green, and that Ireland will free herself as soon as England has war with America. "Not so," answered Mallet, "you will be kept down by 100,000 troops." Haughton rejoices at having Airy in prospect as Pres. R.S. ; says his paper on ' Perturbations of Earth and Venus,' one of the grandest achievements of modern science, far beyond discovery of Neptune. That paper, and his article on * in the ' Encyclopædia Metropolitana,' place him at the head of science.

April 20. Mr. Airy's letter was laid before our Council, and was ordered to be printed in the minutes. It makes an unfavourable impression.

April 22. Our President's last *soirée* well attended. " Lucifer Lowe " one of the visitors.

* The diary is blank as to this, and the article in question is difficult to identify.

The Rt. Honble. Robert Lowe, designated "Lucifer," was so named from his having proposed to increase the revenue by taxing lucifer matches. This caused so much excitement that the proposal was withdrawn, and the income tax was increased instead.

April 23. My sixtieth birthday. How life runs away! To Zoological Gardens and Hampstead; vegetation looks lively from April rains.

April 24. Dr. Sharpey says that Mr. Airy's letter is a mistake; that with a cab to and from Charing Cross and his house, and Greenwich rail, he might solve the difficulty.

April 25. President says Dr. Robinson's letter on Airy's letter is very severe, that he will show it me. That Mr. Stokes dreads Huxley's being President, and so accepts Airy. That once at Lord Rosse's, Airy and Hamilton competed in recitations of poetry, Greek, Latin, and English; that Airy had a prodigious memory, and could recite all Scott's poems, or any part of them taken at random. That he (Airy) was brought up on the farm of his uncle, Biddell, near Bury St. Edmunds.

"Our Club" held their Shakespeare night. Maclure and Walbrook brought Macnee, a Scottish R.A. Opposite me sat Sala, a bronze-faced, bright-eyed, somewhat dissipated-looking man. I sat between Joseph Durham and Moriarty. Sir Julias Benedict in the chair. He made a fair speech, showing how susceptible his writings were of musical treatment. When Benedict left the chair, Hepworth Dixon took it, and Gerstenberg likened him to Shakespeare, whereat Horace Mayhew was wroth. Young Dibdin sang three good songs. Sala made a good speech and sang 'My Maryland' in a way that made me feel as a Southerner. At the end another vexation was the charge of three shillings more for the dinner than was advertised.

April 28. Private view of Royal Academy; my first-sight impressions were Millais' 'Chill October,' Vicat Cole's 'Autumn Gold,' Leslie's ' Nausicaa ' and ' Maids,' Millais' ' Moses,' Leighton's ' Hercules and Death ' and ' Peasants in St. Peter's.'

Dr. Percy called. Says he and his friends will oppose most vehemently the scheme for transferring the School of Mines to Kensington, that the arguments in favour thereof are fallacious, that he persistently refuses to be examined by the Commission. That Huxley's present income is £1000 a year.

April 30. To Buckhurst Hill after luncheon ;

country looking delightfully green and fresh after April rains. Three inches have fallen during the month, 260 tons to the acre.

May 1. After four to the Horticultural Gardens.* The dome of the Albert Hall looks effective beyond the range of Exhibition buildings. Saw the crowds come swarming out, ladies' hats pictures, such as one sees in portraits by Sir Peter Lely and Sir Joshua Reynolds. Went into the Hall, up to gallery, all round, out on balcony, good view over Hyde Park, and thence the Albert Memorial looks well. Went down in lift.

May 2. Sir John Rennie called to say he is going to travel for six weeks. Says Huxley's speech at Royal Academy was in bad taste and very poor. 'Damnbiology' was his word. Says he objurgates the civil engineers, tells them they are not engineers, they get all their work and all their calculations made for them, whereas in the last generation the civil engineers did everything for themselves.

Spoke in praise of Airy, says he will make a genial President, but not to be compared to Sir Edward for grace and manner. Sir Joseph Banks, he says, was a despot, but very hospitable.

* Then in South Kensington.

His migrations from town to his country houses were regular as clock-work. "Sir Joseph was king of the fens, but my father was king of Sir Joseph; he had drained the fens, you know. In that day if you did not get drunk, you were supposed not to do honour to the hospitality of the house. My father once arrived at Revesby, and wishing to depart early next morning, begged the butler to take care that the post-boy did not get drunk. 'I will see that you get off in good time, Mr. Rennie,' answered the butler, 'but I can't undertake to keep your man sober, because the very last time he was here I had to promise him he should get drunk the next time he came.'"

May 6. To the Horticultural. Exhibition Road crowded with vehicles, some for the Exhibition, some for the Gardens. The band played, and the many gay dresses in the sunshine were a pretty sight. The programme, sold at one penny, was printed in the Exhibition, of which it contained scraps of news; a little paper entitled 'The Key.'

To the Meringtons' to dinner. On the way home at 11 p.m. heard the nightingale sing in Kensington Gardens within a few yards of the road.

May 10. President called to talk about the proof of his evidence before the Science Commission, which he has sent me to correct. He does not show sufficiently how much the Royal Society have done for the Government, but appears unwilling to enlarge his statement. The letter written to him by Lockyer, the secretary, was not intended to limit the scope of his answer.

The President spoke again of his wish that Lord Salisbury had been proposed as President, and of letters written to him by Fellows requesting him (Sir Edward) to continue in office. "But I can't," he said. "A President who can't hear what is said when in the chair is of no use. Besides, I want time to finish my magnetism."

He reflects severely on Mr. Airy for his letter asking the Council to allow £100 a year for Presidential expenses.

May 11. At our coming meeting a paper by Dr. Marcet on the Constitution of the Blood. He shows it to be mostly a colloid fluid. In the discussion which followed Dr. Gibson and others denounced beef-tea as void of nutritiveness. With solid food, or the beef powdered, mixed therewith, it is then nutritious.

May 12. A dinner party; Rev. J. Congreve, Mrs. Lynn Linton, Miss Delf, and Dr. and Alice Bird. We had a lively discussion on morals and Christianity.

Mrs. Linton doubts if —— ever lived, sees nothing in his system worthy of belief. She feels without home, and without hope. But she is very agreeable, intellectual, and fascinating.

May 13. Talk with Dr. Sharpey. Contrast of present mode of electing Fellows and the Council with that of former days. Banks and Lord Northampton chose whom they would on the Council; and at a ballot for Fellows, Banks would send Dryander about to say, "The President says you must blackball this man."

Lowe, Chancellor of the Exchequer, once up at the Athenæum; some one objected that he was albino. "I suppose he needs blackballing," said the Dr.

Duke of Sussex once asked Pettigrew about a man who was candidate for F.R.S. Pettigrew answered by mysterious looks and shakes of the head. When the case came before the Council the Duke said, "Oh, that man won't do. He's a ——," so he had interpreted Pettigrew's pantomime. There was a great row in consequence.

May 15. Sir Henry Holland called, desiring

R

to have Sir Ed. Sabine's signature to a memorial praying that Sir John Herschel may be buried in Westminster Abbey.

Dr. W. B. Hawkins tells me his father remembered seeing an old man, who when a boy-had seen Judge Jeffries' entry into Dorchester to hold the Bloody Assizes.

Sir W. Logan says Principal Dawson is an agreeable man, but too speculative, and with an excellent opinion of himself. Sterry Hunt is clever and ingenious, but fickle; once joined the papists, then left them, and is as uncertain in his theology at present as in his science.

M. de Liancourt says there is reason to believe that Italy is plotting against France in the hope to regain possession of Savoy.

May 19. At 11 a.m. to Dean Stanley at the Deanery; he wished my presence at funeral of Sir J. Herschel to identify the F.R.S. and group them in the Sacrarium. Women were sewing the white border on the black cloth round the grave up to within a few minutes of the door being opened. Dean told me about 300 tickets had been given out. Visitors began to enter N. transept door at 11.30. About thirty-five F.R.S. Procession entered at twelve. Our President, the Astronomer Royal, Sir J. Lubbock, Duke de Broglie, Sir H. Holland pall-bearers. Owen with black skull-cap looked very grim. It seemed to me that to a real mourner, the read Service would be more impressive than the full choral Service. The Dead March in Saul to finish with was most appropriate.

May 25. At our evening meeting Mr. Lowe, Chancellor of the Exchequer, was elected a Fellow unanimously, twenty-three votes and not one black ball. Legros Clark, thinking the abstract of his paper on 'Mechanism of Respiration' too short, asked leave to supplement it by an oral exposition. But when called on he could say nothing to the purpose, merely recite observations, and give no results. To Dr. Sharpey's appeal that he should define difference between male and female respiration, he gave a very lame answer.

May 27. Talk with Dr. Sharpey about the destruction and the atrocities now going on. He mentioned the shipwreck of *Medusa* and *Alceste* as a proof of essential difference between French and English character. We agreed that the Thiers Government will be called to account for having contributed to this result.

May 30. Mrs. Linton and Mr. Congreve to

dine; we had an interesting triangular talk. She told me her history; her father a clergyman, she the twelfth of twelve children; mother died six months after birth of twelfth. No education, life a scramble. Left home at eighteen and began to write. Her scepticism and lack of hope are accounted for; but Mr. Congreve's arguments made an impression.

May 31. To Reading; reached home again at 11.15. Among the party at John's was a Mr. Smith, whom they called the Archbishop, a jolly-looking person: the convalescent nephew looks like a boiled brigand, and the sight of Theodore's two little boys made me feel how fast the young generation is pushing the elders out of the world.

June 3. To Haslemere; afternoon with Mr. Simmons to Marley, where Mr. Pratt is building a house in command of the glorious prospect; down through copses to open meadows, across a moat and to Shulbred farm, an old monkery, vaulted basement, new cellar and dairy, refectory above, remains of old frescoes.

June 9. To the Opera Comique with Cousin Harry: the company of the Comedie Française, 'Il faut qu'une porte soit ouverte ou fermée,' 'Valerie,' and 'Le Médecin malgré lui.' The theatre, the prettiest in London, the acting excellent. M. Got as Sganarelle was admirable, falling back in chair excepted.

June 15. Last meeting of our session. R. Lowe, Chancellor of the Exchequer, came to be admitted with nine of the newly-elected candidates.

June 17. Called on Mrs. Linton. Talk about social morality. She is grieved and vexed that she cannot see a way out of present chaos. Might as well be grieved because she can't find out the occasion or cause of gravity, or of terrestrial magnetism.

June 18. To Balham by rail, then to Macmillan. Walk to Tooting Common, glorious day, lively breeze, low masses of cloud, pretty lane from Doulton's to Congreve's. Doulton's new house out of harmony with the old of which it forms part. Heard the cuckoo among Macmillan's trees. At dinner was Mr. Keltie, formerly an assistant editor of ' Chambers'.' Afterwards Lloyd-Dawkins called.

June 20. President called with Lady Sabine. Says he has heard that Mr. Airy intends to apply to Government for table-expenses, because so many foreigners call on him. And that he (Sir Edw.) had overheard some one say, "The

Royal Society is becoming a refuge for people who want to make a little money."

The President having to attend the Queen's breakfast on Friday, will not come to the Dredging Committee. He thinks Dr. Carpenter looks too much to self; that to push the question forward because the Americans may send out a dredging expedition is an unworthy motive, and that it may be left in abeyance at all events till November.

Macmillan tells : When 'Bentley's Magazine' was projected the title proposed was 'Wit's Miscellany'; after discussion some one said, why not call it the proprietor's (Bentley's) Miscellany ? "Pray don't go from one extreme to the other," retorted Jerrold.

June 21. Arrived at W. Crookes', Merington Road, where Home the Spiritualist was holding a *séance;* four ladies, six men present. Before my arrival, Home by putting his fingers into water at one end of a board had caused the other end to descend. Home is somewhat sallow complexion, light brown or sandy hair and moustache, high cheek - bones, dreamy-looking eyes. He told of his experience of second sight. By and by I was asked to place my hand on the table to feel a tremor. There

was a tremor visible in the vessel of water,
occasioned no doubt by passing trains on the
railway, but the company thought it due to
occult influence. I departed at 10.20.

June 26. Dr. Sharpey called. He has been
on a visit to Mr. Prestwich; and met Mr. Grove
and had much talk with him, and says he is
sorry that Mr. G. was not proposed as President
in place of Mr. Airy.

The Doctor ought to have called on Mr. Grove
last November when it was suggested to him,
and he promised to call, and afterwards changed
his mind.

July 5. To garden-party at Newmarch's, Clapham : very gay and pleasant, and the weather
turned to fine. Met the Masons, Baldwin-
Brown, Sopwith, Chadwick, and other old
friends.

To dine at Dr. Bird's ; one of the guests was
Morton, author of 'Box and Cox.' Ionides
offered to give me letters of introduction to
Constantinople.

July 7. Dr. Sharpey tells me of Mr. Grote's
bequest of £6000 to endow Chair of Moral
Philosophy in University College. If a clergyman is ever chosen Professor, the pay is to stop
and accumulate for his successor. Cornwall

Lewis once said to Grote, "You are a bigoted unbeliever."

July 10. Dined at the Albion by invitation of Literary Committee of Corporation. Mr. Todd in chair; about seventy guests. I sat by head table, and had next me Mr. Scott, a clever Scot and Stockbroker. Literature, Science, and Art were proposed. Hepworth Dixon and my name for the first; Odling second; E. M. Ward third. Dixon made the speech for literature, whereby I was spared a painful effort.

July 11. To Tooting, dined, then to Club meeting at James Rose's. Mr. Congreve ill in bed.

July 19. To dine at Mr. C. Mason's, Denmark Hill. Baldwin Brown, Mr. Abbott of City of London School, Mr. Payne, Miss Buss, and others present.

July 21. At noon the Emperor of Brazil, attended by Dr. Hooker and Mr. Hanbury, came to look at the Royal Society Rooms, curiosities, and portraits. He has an imposing presence, but the voice of a cockatoo. Of the things he looked at, our Charter Book interested him most. The royal signatures he treated with indifference, but looked eagerly at those of the early Fellows, and Newton, Wallis, Flamsteed, and others, showing that he knew the names of the best men of successive periods.

August 1. First day of holiday. Left house, slept at Namur, thence down the Moselle, on to Paneveggio, down to Verona. Kreuznach on way back, and returned.

September 15. Found all well.

October 8. To Sheen Lodge. Walk and talk with Professor Owen and Mr. Chadwick. Dine with the Professor. He read extracts from Morris' poetry, and said that in William Morris we have one to fill the gap which will one day be left by Tennyson.

November 10. To private view of Aquarium in Crystal Palace.

November 22. Treasurer audited my accounts, and found them correct in every particular.

December 14. Judge Richardson of the Treasury, Washington, to dine. He thinks Englishmen uncommonly good fellows when thoroughly known.

December 20. Mr. Knight from New South Wales, the clergyman whose name my son Walter has often mentioned in his letters, called. He tells me that Walter's entrance into the Church and undertaking pastoral duty arises from an earnest and serious conviction,

1872.

January 2. Mr. Simmons of Haslemere called. He is now one of the Surrey bench of magistrates. Lord Lovelace (so he says), having a hearty dislike to Liberals, long opposed his appointment.

January 2. Dr. Sharpey tells me that Huxley willingly entertains the notion of being Secretary R.S.

In a subsequent talk Dr. Odling expressed approval.

January 8. Duplicates cleared from Library.

January 10. To dine at Doulton's, Tooting. Masson (of Harrow), Newmarch, Dr. Perry, Macmillan.

January 16. To Club. Ashurst, Norton, Dixon, Harding, Captain De Rohan, phallic Simpson. Hamber on Karagoose.

January 20. Dr. Sharpey called. Gave me some particulars of his autobiography. How hard he worked in Berlin even on Sundays, and how his preparations were shown by the Professors as models of workmanship to the other students.

January 24. Legros Clark called. Complained that his paper had not had fair play. I set before him the truth of the matter, and made him see that pains had been taken to place opportunities in his way, but that he had neglected them all. He went away with easy mind.

January 31. To Club meeting. R. Cooke in chair, Dixon, Verey, Hazlitt, O'Connell, Cosens, Lloyd, McClure.

February 4. To Coleherne Terrace. Dine at Miles'. Mr. and Mrs. Frank Miles present.

February 5. Called on Mr. Curling about an infirmity which he pronounced to be a bad case of varicocele; good health has been my lot through life, and I must not repine now.

February 12. Jarrolds accept terms, and will now publish ballad 'Prisoner and his Dream.'

February 13. To Club: Cunningham, Moriarty, Ward, Lawrence, Norton.

February 16. Called on Wesley. The Smithsonian Institute now allow him £50 a year for his services as agent. Saw preparation of seats for Thanksgiving Day on site of Law Courts.

February 26. Mr. Darwin called to talk concerning the rumoured nomination of Huxley as Secretary; he will support it.

March 4. To music party at Royston Pigott's.

Our Secretary and Treasurer agreed that we should close at four on Saturdays.

March 14. Posted proof of Ballad to Jarrold's. Saw William Chambers in the Row—wasted £40,000 of their property.

March 18. Hepworth Dixon and Sylvester called. S. is a candidate for the School Board vacancy in Marylebone.

March 28. Evening to Banbury for Easter holiday.

April 3. To London. A greybeard with his denunciations of Society bored me. I changed compartments and had good talk with a Queen's messenger.

April 23. Letter from son Harry after four years' silence. He is now a petty officer on United States ship in Pacific.

April 29. To Mrs. Linton's At Home. Saw sundry friends, and Mrs. Dickens, Mrs. Dallas and Ronniger. Talk about women's rights.

May 14. N. Lockyer in chair. Woolner was proposed; I seconded him.

May 31. Club. Cunningham in chair; Norton, Hamber, Col. Grant, Jeaffreson; Norton says all our College education is rotten.

May 28. To Civil Engineers' *soirée* at the International Exhibition.

June 12. G. W. Prime, the American, who was my travelling companion in Norway, called. How glad I was to see him looking so hearty! He is come to take a tour with his wife. At 10 p.m. music at Johnston's.

June 13. One hundred and twenty friends came to my At Home. Willis supplied refreshments, and all went off well.

June 2. Mrs. Linton's At Home. Talk with Hazlitt and Leland. The latter is studying the language of Gipsies.

June 24. To see M'Callum's Egyptian pictures. One a very desolate view of the Desert.

June 29. *Soirée* at College of Physicians. Dr. Beale delivered himself of a growl on the prospect of Huxley being Secretary.

July 1. Mr. Busk and Mr. Savory express their full satisfaction at the nomination of Huxley.

July 2. Conversazione at Royal Academy; stayed till midnight.

July 5. To dine at The Ship, Greenwich. Dinner given by Clowes, stationer. Clowes and Son, John Murray, Dr. W. Smith, H. Cole, Murray, jun., R. Cooke, Shirley Brooks, Crowdy : good talk and good Rudesheim.

July 6. To Forest. Met Miss Huss, Chessar. Stayed till Monday, but in much pain.

July 9. To survey Acton Estate with Treasurer. To Arthur Tooth, to buy pictures with G. Palmer. He bought a Linnell and Lucas.

July 12. Called on Mr. Curling. He advises an operation as the only cure for my pains—torture.

July 13. At 3.45 p.m. came Curling and M'Arthur. I was five minutes under the knife and winced at the pain, but did not kick nor howl. Then to bed and hot brandy-and-water.

July 14—17. In bed. Dr. Keats, M'Arthur, Treasurer, Merington, Rennie, etc., called.

July 18. Out of bed half-an-hour. Giddy and weak.

July 12. Up two hours.

July 20. Up all day.

July 23. In office. James Rose called. At Club Meeting at Doulton's; subject, Wordsworth and Tennyson; not a vote was given for Tennyson. He spoke of Swinburne's pamphlet *Under the Microscope.*

August 1. To Ostend; there till 6th. Then Brussels, Heidelberg. Then away through Munich, Innspruck to Paneveggio. Thence Pusterthal and Drauthal to Klagenfürt, St. Veit, Murgthal, Brück, Vienna, Lindau, Luzern, Basel, Paris, and home September 15.

September 28. Maxwell Simpson tells me he is now Professor of Chemistry in Queen's College, Cork; that he likes London best as a residence, that Tyndall is going to give a course of lectures in America, where he will "like the lionising." That Debus, Armstrong and others are candidates for Odling's vacant place at Royal Institution.

September 21. Dr. Hoskins called. He hopes to get his collection, which has been favourably noticed in the State Papers, published by the Camden Society. I'm glad that Huxley is to be our new Secretary.

H. W. Bates says with reference to the quarrel between the Geog. Soc. and Lieu. Dawson, that Livingstone is a man of hard nature, energetic, and self-reliant, but bitter in mind. That the expurgations in a letter complained of by L. were omissions of bitter remarks about Arrowsmith, against whom he has great animosity; that the map engraved by Arrowsmith for L.'s book cost £400.

Bates says the Geographical map-room is scarcely made use of by the public, and that the Geog. Council are anxious to stimulate interest therein.

R. Mallet says Dr. Percy's book on iron is nothing like so good as reputed, that the chapter on molecular constitution of iron is pillaged from Mallet. That Percy can talk loud, and frighten timid folk; but is afraid of those who can answer him. That Graham always referred to his investigations of the molecular constitution of iron as the best ever made.

Dr. Stenhouse says Faraday was selfish and narrow-minded. That a man once went to him, as he himself had gone to Davy, and that F. sent the young man to Graham, of which incident Graham made a standing joke.

September 23. Dr. Sharpey home from Tyrol; says Godwin Austin told him good coal was now dug in the Boulognais. That he (Dr. S.) does not believe what is said about Dr. Kirk's jealousy of Livingstone; neither do I.

1873.

May 2. Private view at Royal Academy;
went in for an hour in forenoon and up to six
afternoon. No startling pictures, but many very
pleasing. Canon Liddon portrait good and
singularly intellectual in expression. Mr. Spottiswoode by Watts too dark. Talk with C. Barry,
who says new house will be ready by September 1.

Talk with J. Marshall, who is anxious to get
the lectureship at the Academy in succession to
Partridge. The emolument is £100 a year; he
would give up his post at South Kensington
(worth more) to take it, because of the distinc-
tion it gives. He wishes me to speak a word in
his favour to any one who had influence. We
spoke of Dr. Sharpey, whose eye is to be
operated on this day week. Marshall devoutly
hopes the operation may succeed. I questioned
as to the Dr.'s ways and means. M. knows
nothing; but being a trustee under the Dr.'s will
(jointly with Robson of the University), he
expects to know something shortly. Thinks the

s

resources are small, and knows a few men who are willing to contribute among them £100 a year. The Dr. was offered the interest of the Sharpey fund at the University, but declined it.

Dr. Stenhouse called to enquire whether James Young was one of the selected candidates. He is still angry with Lowe for abolishing his post as assayer, which produced him £600 a year. A compensation of £500 was granted him.

1874.

January 1. Glorious bright day, same mild temperature that has prevailed throughout December. To Dr. Bird, who says my giddiness arises from sleeping after the first waking in the morning. Says that Merington's case is hopeless. I always feared he had put off too long seeking good advice.

Mr. Bentham says he heard of Sir G. Airy's rejoinder to Owen : thinks it best that Owen's should not be noticed. Says Kippert is good accountant, but slow-coach in all besides: that the Linnean Tracts have never been sorted, and that he (B.) is now sorting. That Kippert was elected by the Society in opposition to the Council. Mrs. H. called for advice : her husband keeps her entirely in the dark about his business and money matters. Jas. Young called with four ladies, three of them his daughters, not good-looking ; the other, Dr. Livingstone's daughter, is a bright-looking Scottish lassie with downy lip and black eyes.

The Academy scaffolding cleared away and levelling of court-yard begun; the new façade is a failure.

January 2. To City, rain and gloom. Paid £250 on account of Catalogue to Paymaster-General at Bank of England. This makes £1000 repaid. Mr. Huxley called: spoke of Sir G. Airy's rejoinder to Owen as being a clumsy mistake; that Mr. Stokes settled it by writing to Sir G. a wonderfully clever diplomatic letter.

January 3. Clearing off correspondence. President called with Swedish damsel. Was glad to hear that Sir G. Airy's letter had been so cleverly disposed of. To Kensington to dine. Talk with Mr. M. about Harry, and told him that Bird considered his case "very serious." "My own instinct tells me the same," he answered sadly.

January 5. To the Treasurer; agrees it will be best to read Dr. Tyndall's paper on the 15th. Called on the Dr.; he agrees. But he cannot be at our meeting on 15th, as he needs a whole night's sleep to prepare him for his lecture at the Institution on the 16th.

January 6. Had time to sort away the temporary parcels on my office-shelves, and to plan Newton collection for glass-case. James Rose called. As a whole he thinks the Landseer collection unsatisfactory; animals too human, men too animal. To "Our Club," only five present. Coleman in chair, full of anecdote as usual. The Arlington Club is a sporting club; meets in Arlington Street. If a member makes a bet knowing he can't pay, they eject him at once, and his name is struck off the lists at Tattersalls'.

January 7. Tupper called; says he has been reading his own works to large audiences in Scotland; that he believes in spiritualism, is to meet Crookes, and at Carter Hall's saw a large dining-table rise 2 ft. from the floor. Mrs. Linton called. Chatto and Windus to give her £550 for a novel which she wrote in eight months. Mrs. Thomas, Lilian and Betsey Burton to tea; talk about art. They agree with me that your true artist invents and does not need to copy. Do not like Holman Hunt's picture: but do like Doré's 'Christ leaving the Prætorium.'

Sir Edward and Lady Sabine called. His portrait is to be painted by Watts, for the regimental room at Woolwich.

January 8. Talk with Col. Strange at evening meeting. He says that Dr. Pritchard's letter in 'Times' about origin of Observatory at Oxford is a lie; that Pritchard dislikes Lockyer, fearing that he aims at being director of the new Observatory.

January 10. Talk to Mr. Salter about B.'s visit, and his impression that he had never taken any salary while Secretary. The same prevails (said S.) in all his statements about money-matters; speaks of his donations as double the actual amount. Salter paid him £13,000 during the years of the Secretaryship, proceeds of the business in Broad Street, and had but clerk's pay; and yet B. thinks the obligation the other way. That in private, his temper is passionate; that he raged on going home when the Scientific Relief Trust was accepted by the Council; that he came to London at Xmas only to avoid hearing at Selborne the first sermon of a curate he disliked.

January 10. Further Mr. Salter said that the majority of F.R.S. would prefer a nobleman as President.

January 13. Our Club. Eastwick in chair. Two Napiers visitors: one in the Temple, other diplomatist. Says when Gortchakoff retires Ignatieff will most likely succeed.

January 15. Council Meeting; Tyndall on experiments to make noises in foggy weather at sea, and on behaviour of atmosphere, at evening meeting, which was crowded.

January 16. Mr. B—— from Australia to dine. Talk about the Colony and son Walter. Lying and tippling he says were characteristics of the colonists.

January 17. To dine at Gardner's, St. John's Wood Park. Dr. Boycott, Mr. Birch artist, Dixon, Dr. Richardson. The latter sang 'The Blacksmith,' a good song, and told good stories.

January 18. Dine at Thomases'. Talk of life, death, and immortality. The two young men and Miss B. of opinion there is no immortality.

January 20. Our Club, quiet party, no chairman. Watts told about Wentworth, the Sydney politician; father was governor of Norfolk Island; mother one of the convicts; wild but eloquent. Wrote letter to Lowe, pleaded that Lowe's habits rendered thereby a breach of the peace impossible. Dixon talked of the completion of his two Queens, and of rage of priestly party in Spain at stripping of veil from the great Queen Isabella.

January 22. Two Committees. Library at four. Privileged classes, 5.30. At evening meeting; Brunton, and Fayrer's paper on Snake-poison after. Dr. Aquila Smith said a good sign of the

times was that a priest had been pelted with
mud in Limerick; that the Protestant clergy
were greedy of power; that the country was
rich, there being £28,000,000 on deposits; that
Home Rule was humbug; that at the Union
many Irish expressed themselves happy at
having a country to sell, seeing they got so good
a price for it.

January 23. Prof. Newton says it came out
through the ladies that Prof. Miller resigned
because he did not like Dr. Hooker as President; that many Fellows think the same, and
that a large party is in favour of the Duke of
Devonshire.

January 26. To Gray's Inn Road, met
Harry, bought outfit. Letter concerning him
from J. Hamilton, New York. Disheartening.
M. Merington to dinner. Told me of Margaret's
enjoyment at Berlin, of the proceedings with
Princess Frederick Charles, interview, with-
drawal, stipulations, visit of Hofrath. Fifty
pounds paid and handed to the Governesses'
Institution.

Dr. Stenhouse maintains that F.R.S. should
have nothing to pay; that scientific men are
more at a disadvantage than any others in the
community.

1877.

October 23. C. called full of a grievance and
morbid withal. He has heard from Wallace
that Prof. Stokes and three other Fellows have
stated in writing that he (C.) is wrong in his
controversy with Carpenter, and Carpenter is
right. That Carpenter sought to trap him with
a proof which he (C.) declined to correct; that
Carpenter has acknowledged himself to be the
author of that hostile article in the ' Quarterly '
a few years ago; that Carpenter has written to
him (C.) asking, "Why do you never praise me ?"
and so C. is unhappy, can't settle to work, says
his self-respect will compel him to answer any
further attacks or charges. I advised him to
hold his peace, or rather his pen, to decline to
believe what Wallace had written, to remember
that vanity was very often mistaken for self-
respect, that silence and honest work, work for
the love of it, are stepping-stones to great-
ness.

October 25. First Council for new Session.

Asked Mr. Stokes if he had written to Carpenter that he thought C. wrong. " No. All I wrote was, that C. put forward no theory, but seemed to lean to the idea that the movements were due to the direct action of light." So much for epistolary gossip.

Mr. Marshall to speak about Downes' paper. Told me of Dr. Sayre's operation for cure of crooked spine. Patient is suspended by chin and back of head. Weight of pelvis straightens the column, plaster of Paris belt put on, and in six months the cure is complete.

October 27. Dr. Huxley called ; wrote letter of thanks to John Evans for gift of Herschel portrait; told him he should have it on foolscap if he wished ; but that two augurs knew the value of such display.

Wrote also letter to Admiralty recommending them to send the *Valorous* duplicates direct to British Museum.

Spoke of Dr. Baxter Langley and Swindle-hurst. When the Sunday Lecture Society was at work, they got notice that public lectures, not religious, were contrary to law. Langley thereupon, without consulting any one, went and registered it as Religious Recreation Society. Huxley withdrew in disgust; one of the objects of the Society had been to fight the law, and that was set aside by the foolish registration.

Swindlehurst wrote to him repeatedly to beg for lectures to be delivered to working classes. Would take no denial, and being full of zeal informed him (Huxley) that he was neglecting a duty higher than any other. To this an answer was written which gave Swindlehurst his *quietus*. "Enthusiasm in business," said Mr. H., "is to me very suspicious."

October 28. Called on Dr. Stenhouse.

1878.

March 14. At our coming meeting, G. H.
Darwin's papers on 'Haughton's Estimate of
Geological Time' was read. John Evans spoke
thereon fluently and well, and agreed with
Darwin. Tomlinson's paper on 'Super-saturated
Solutions' brought up Dr. Carpenter, who thought
that Tomlinson might study the formation of
ground-ice in the sea, and he spoke of layers
of still water deep down of a temperature of
23°, which freeze if agitated. Shoals of fish
dash therein and rise to the surface coated with
ice as seen on coast of Norway. Dewar and
Wyville Thomson had hearty laugh at this over
their toddy.

April 8. Mr. Spottiswoode told me he had
seen Clifford off for Gibraltar, and he has hope
of his recovery.

December 1. Death of G. H. Lewes announced
in 'The Times.'

December 5. Crowded evening meeting.
Crookes' paper on 'Illumination of Molecules.'

His exposition was twice as long as it ought to have been, and too full of detail. To describe what you have done is of more importance than how it was done, seeing that here the hearers have either the how in their hands, or will soon have it in Proceedings. Crookes' experiments were clever and conclusive. In the discussion Varley told what he did at the negative pole ten years ago; De la Rue commended: Prof. Stokes said that the light in the glass globes and cylinders was due to impact, but hard to explain. His suggestion of outer film shrewd and profound. Tyndall, his recent researches in natural history had led him away from such experiments as those before the meeting; it seemed as if Crookes had repeated and extended Bernouille's experiments, but what was the cause of the light? Dr. Carpenter mentioned Faraday's rotation of a ray by magnetism. Prof. Stokes spoke again, and Crookes replied.

Within the first half-hour, Mr. Justice Grove and others, weary with the details, went away. The meeting occupied two hours. In subsequent talk President, Prof. Huxley and Mr. Hulke agreed with me that the occasion was spoilt by detail: which I mentioned to Prof. Stokes and Mr. Crookes.

December 6. Prof. Dewar says a scientific man should never exceed twenty minutes in an exposition; that he knows by experience that much may be spoken in that time. That he is shocked when in London by the self-seeking of scientific men; no man caring to work for love of the work.

1879.

January 3. Talk with President (Spottis-woode) about the details in reading papers at evening meetings. Of the necessity for sum-marising and making the meaning apparent to unscientific hearers. Mr. Huxley's treatment of the papers, December 19, was praised as masterly. The new arrangement of using up papers quickly will favour this.

January 4. Met Dr. Percy at private view of Academy, Winter Exhibition. Some years ago he introduced L—— to 'The Times,' and an article of three columns was published. Delane sent a cheque to Percy to be handed to L——, unless he (L.) thought himself paid by the much praise he had bestowed on himself.

October. Dr. Carpenter says the meeting of Brit. Asso. at Sheffield was not good. That Huxley seemed sore under mention of Bathybius ; that —— of the *Challenger* can make Bathybius by mixing spirits of wine with sea-water ; that Dana holds that the great sea-basins have undergone but little change, are so to speak perpetual; that Abbé Beynard (Belgium) holds that metals are still forming, which, if true, will tend to modify existing views as to the structure of the earth.

Says working men of Sheffield are intelligent and ask shrewd questions after a lecture; that before the spectroscope had been used to test the Bessemer flame, one of the men queried as to whether it might not be so applied.

Formerly little masters had small mills in all valleys round about. Now great factories have taken their place; and the residences of the great masters are palatial.

Wyville Thomson has been seriously ill.

Dr. Duncan hired a house in outskirts of Sheffield during the British Association Meeting. Says that Allman, the President, was named King Lear; that his address was retrograde biology; that Huxley's face was a wonder to behold when Bathybius was mentioned. That the Abbé Beynard is a joker and smoker though a Jesuit; that while travelling in Scotland heard driver call refractory horse a Jesuit. Asked if knew what it meant, "Na! but 'twas something worse than 'deil!'"

At Charterhouse School, Archbishop of York

contended for history and classics. Husk answered, "All very well, they cultivate the memory; but with natural science we wish to give them a reasoning faculty." Dr. Duncan is not in favour of teaching biology to boys, it would unsettle their basis of belief.

Duncan edits Cassell's 'Nat. Hist.' Got Parker to write article on ——. Parker being very busy set his son to write. Duncan did not read it. It was printed and published, and then the reviewers found and proclaimed unpardonable errors.

T

1881.

September 27. Mr. Hind spoke of the appointment of Christie as Astronomer Royal, Greenwich Observatory; that it was arranged by Sir G. Grey and three others, one of them being P.R.S. That Henry Smith was asked for his sanction and declined. That Dunkin felt aggrieved and wanted to retire, but was persuaded to stay as Chief Assistant to help Christie for two or three years. That the new Second Asst. is to be of Cambridge and skilled in mathematics. That ——— of Oxford put himself out of court by paragraphs in the 'Standard' and 'Daily News' suggesting that he should be the new Astronomer Royal.

1882.

March 2. A large afternoon meeting, for the news had spread that the Prince of Wales was coming to be admitted. At 4.20 Mr. Gladstone arrived, looking hale and self-possessed. Just as the President took the Chair the Prince arrived. Mr. Gladstone undertook to introduce him to the President. The Charter Book lay ready. I placed the pen in the Prince's hand and pointed to the place where he was to sign. " I ought to have done this long ago," he said, which was true, for he was elected in 1863. Then he seated himself in Dr. Foster's chair, and I fetched a chair on which the Dr. sat looking very much out of place. The Prince should have sat on the front bench among the Fellows. Dr. Foster read seven certificates ; Mr. Stokes read the list of fifty-one Candidates for the Fellowship ; then Mr. Huxley gave a masterly exposition of his paper on the ' Salmon Disease.' He ended by saying that supposing a cure could be found, under present circumstances, the cost of the cure would be as great as the loss by disease. The occasion of the disease is a vegetative parasite.

Then Mr. Siemens expounded his 'Conservation of Solar Energy,' his argument being that all the heat or energy which the sun gives forth is returned to the sun. The hour was then 5.50. The Prince having other engagements rose; made a neat speech, hoped that the International Fisheries Exhibition to be held in London in 1883 would assist in the further clearing of Mr. Huxley's subject. As regards Mr. Siemens' profound speculations he (the Prince) had not knowledge enough to judge of them; but he had no doubt they would provoke much scientific discussion. He then withdrew, attended to the door by the President, Treasurer, and Secs. Mr. Gladstone walked away unattended before five o'clock.

Sir William Grove began the discussion by stating that forty years ago he put forth the notion of an attenuated form of matter throughout space (Correlation of Physical Forces); that the theory of ether and undulations is one that can be adapted to any phenomena, and phenomena to it; that "disassociation" was an inappropriate term; that substances could be resolved to any extent, given a wide range of temperature ; that disassociation implied combination.

Mr. Stokes looking very grave ; news of the death of Rev. Dr. Robinson came to him yesterday.

1884.

February 22. Dr. P. M. Duncan says that Dr. Acland of Oxford has drawn his salary for years, but has not given lectures.

Dr. Gilbert, who has just been appointed Sibthorpian Professor of Rural Economy at Oxford, expressed a wish that it could have been put off for a year, he being so overdone with work. Sir Joseph Hooker answered, "Oh, you must adapt yourself to the spirit of the place, and give the lectures only when you find it convenient."

October 13. Dr. Carpenter says Mr. Huxley is seriously ill, and has been ordered a three months' holiday by Sir Andrew Clark. Is to start for Venice this week; stay a month, then home to marriage of his third daughter; then away for two months with his wife. Mrs. Collier, Mr. Huxley's daughter, is, so says Dr. Carpenter, very delicate, so that her friends are anxious about her. Her picture, two girls card-playing, was accepted by the Royal Academy without question. The players are portraits of her two sisters. Sir Wm. Siemens bought the picture. After his decease Lady Siemens gave it as a present to Mrs. Huxley.

October 14. Dr. Pettigrew called. He persists in believing that his flying machine will be successful. He has suggestions from torpedo-makers as to how a boiler should be made to resist the pressure of 600 lbs. to the inch. The copper tubes are drawn out of the solid metal. With this the delay between the rise and fall of the wings will be overcome and a gale of wind will not be feared.

With this entry the diary of Walter White leaves off rather than ends. There is little more to be said. Walter White retired from the service of the Royal Society in 1885, in consequence of a rheumatic affection of the right hand, which rendered penmanship difficult, and which is reflected in the manuscript of his journals. As a mark of appreciation, the Council of the Royal Society continued the payment of his full salary to the end of his life. On leaving Burlington House he resided at Brixton, and retained full possession of his mental faculties until close on his death. He died on July 18, 1893, having reached his eighty-second year. His character is apparent from these selections from his memoranda. To one side of it only has reference been omitted. He was much in the habit of writing poems. A volume of them was published as an "author's book" by Messrs. Macmillan & Co. in 1873. In addition to this volume, Walter White's chief publications were:

To Switzerland and Back.
A Londoner's Walk to the Land's End and a Trip to the Scilly Isles.
On Foot through the Tyrol.
A July Holiday in Saxony, Bohemia, and Silesia.
A Month in Yorkshire.
Northumberland and the Border.
All round the Wrekin.
Eastern England from the Thames to the Humber.

Walter White was a voluminous correspondent; but for the purposes of this volume it has been thought sufficient to allow his Diary to tell its own story almost exclusively.

INDEX

THE END

RICHARD CLAY & SONS, LIMITED,
LONDON & BUNGAY.